SpringerBriefs in Materials

The SpringerBriefs Series in Materials presents highly relevant, concise monographs on a wide range of topics covering fundamental advances and new applications in the field. Areas of interest include topical information on innovative, structural and functional materials and composites as well as fundamental principles, physical properties, materials theory and design. SpringerBriefs present succinct summaries of cutting-edge research and practical applications across a wide spectrum of fields. Featuring compact volumes of 50 to 125 pages, the series covers a range of content from professional to academic. Typical topics might include

- A timely report of state-of-the art analytical techniques
- A bridge between new research results, as published in journal articles, and a contextual literature review
- A snapshot of a hot or emerging topic
- An in-depth case study or clinical example
- A presentation of core concepts that students must understand in order to make independent contributions

Briefs are characterized by fast, global electronic dissemination, standard publishing contracts, standardized manuscript preparation and formatting guidelines, and expedited production schedules.

More information about this series at https://link.springer.com/bookseries/10111

Enakshi Bhattacharya

Biosensing with Silicon

Fabrication and Miniaturization
of Electrochemical and Cantilever Sensors

 Springer

Enakshi Bhattacharya
Department of Electrical Engineering
Indian Institute of Technology Madras
Chennai, India

ISSN 2192-1091   ISSN 2192-1105 (electronic)
SpringerBriefs in Materials
ISBN 978-3-030-92713-4   ISBN 978-3-030-92714-1 (eBook)
https://doi.org/10.1007/978-3-030-92714-1

This Springer imprint is published by the registered company Springer Nature Switzerland AG
The registered company address is: Gewerbestrasse 11, 6330 Cham, Switzerland

To my parents,
Tripti and Indu Prakash Bhattacharya
from their third daughter

Preface

With the need for point of care diagnostics, biosensors with high sensitivity and the ability to give results rapidly have become much sought after. This dual expectation of accuracy as well as speed from biosensors is contradictory. In practice, there has to be a trade-off between the rapid results for a quick diagnosis and very sensitive measurements for low sample concentrations. Even if the latter takes a little longer, if the time taken is still less compared to current techniques, it can make a difference in the early diagnosis of swiftly progressing diseases. This monograph describes two biosensors made from silicon, an electrochemical one and a mechanical resonant cantilever that use highly selective enzymatic hydrolysis for the estimation of triglycerides and urea. Well-established standard microelectronics processing and micromachining technology for silicon were used to fabricate and miniaturise the sensors with protocols for the immobilisation of the enzyme on the sensor surface. Careful measurement methods, including the sensor calibration, were developed giving results that agree well with those from clinical laboratories.

Given the multidisciplinary nature of the field of Biosensors, each of the areas mentioned above have been touched upon in this monograph with the belief that it will help graduate students and researchers working in these areas. Since the chapters start from the basics, hopefully, the book will cater to readers with different backgrounds like chemical, mechanical, electronics, physics and biochemistry—the boundaries are anyway becoming fuzzier with time. There is much commonality in the basic principles for developing biosensors even if the biosensors, or the detected bioanalytes, differ. It is a very exciting area with much promise, and the one requirement to pursue it would be an open mind to learn new things as one goes along.

This work has been funded at various times by different agencies listed chrono-logically: Department of Science and Technology (DST); Interdisciplinary Research Programme (IDRP), IIT Madras; Department of Biotechnology (DBT); National Programme on Micro and Smart Materials and Systems (NPMASS); the Centre for NEMS and Nanphotonics, IIT Madras funded by the Ministry of Electronics and Information Technology (MeitY).

Chennai, India   Enakshi Bhattacharya

Acknowledgements

This work, carried out over two decades, has the contributions of several Masters (M.S.) and Ph.D. students from the Department of Electrical Engineering (EE), IIT Madras. I am grateful to past M.S. students Ravikumar Reddy, Indranil Basu, Arun Mathew, V. Hareesh and Gaurav Kathel and past Ph.D. students Renny E. Fernandez and Mohanasundaram Veeramani. Contributions to this work from M. V. Chinnamani, S. Gayathri, Noel Prashanth, Dr. Priyanka Bhadra and Dr. V. T. Fidal from the Centre for Nanoelectromechanical Systems (NEMS) and Nanophotonics (CNNP) and processing support from C. Rajendran, J. Prakash, G. Venkateswaran and T. Sridhar from the Microelectronics and MEMS laboratory, EE Department, are gratefully acknowledged.

My work on biosensors and then bioMEMS was initiated and mostly done in collaboration with Prof. Anju Chadha from the Department of Biotechnology, IIT Madras. Without her, I would not have ventured to add the 'Bio' to my 'Sensors', and I am grateful for the essential biochemistry she has taught me. The readout circuit research was mainly the work of Prof. Shanthi Pavan. EE. Professor Anil Prabhakar, also from EE, is my 'go to' person for cleaning up measurements and data analysis.

I am also grateful to Dr. V. T. Fidal for much help including the careful bibliography. Thanks are due to Vani Mayakannan for chapter formatting and M. Muthaiah for solving all software glitches. Thank you to my present M.S. students Anweshan Chakrabarti, Kaustav Dutta, Abhishek Kumar, and Ph.D. student Sourodeep Roy who helped with figures and references.

I must thank Springer Series Editors Prof. B. Ananthanarayan and Dr. Lisa Scalone for their perfectly timed request to contribute to SpringerBriefs when I had a sabbatical at hand and a pandemic outside. Writing this book filled up much of my sheltered time, though it took painfully longer than I had first anticipated. I am thankful for the timely long-distance nudges from my sisters and nieces as well as for the (unfriendly) queries on the status of my writing from friends, who shall remain unnamed, both off and on campus.

October 2021   Enakshi Bhattacharya

Contents

Chapter 1
Introduction

A brief introduction to biosensors is followed by a consideration of the viability of silicon as a material for biosensors. Electrochemical and microcantilever biosensors are then discussed. The importance of estimating triglycerides and urea, with an overview of the different kinds of sensors currently used, is given. The chapter ends with a glimpse of what is subsequently presented in this monograph.

There are several comprehensive review articles (Kawamura and Miyata [1] and books on biosensors [2, 3]. What emerges from the published literature is that this is a multidisciplinary area requiring inputs from material science, biochemistry, device physics, microfabrication and circuit design areas. A combination of all, or some, of these can lead to sensitive and selective miniaturized biosensors for effective use in the food and health industries. Our work has mainly been on biosensors made from silicon, specifically an electrochemical one and a mechanical one. Starting from understanding the biochemical reaction behind the sensor and optimizing the sensor structure for high sensitivity and specificity involves many technological challenges. Miniaturisation of the sensor, in order to further improve the sensitivity and/or the throughput—among other advantages, brings up yet another set of new challenges, hitherto unseen in the macro devices. The idea of this monograph is to bring out these challenges and their mitigation, taking these two sensors as prototypes.

1.1 Biosensor

The first commercial biosensor was a glucose sensor introduced by Yellow Springs Instruments in 1975 [4]. The concept of an enzyme based biosensor was earlier introduced by Leland C Clark Jr in 1962 and his invention, known as the Clark electrode, measured oxygen concentration in liquids with the help of enzymes [5]. Biosensors are analytical devices which convert the signals from the biological domain into another domain which could be optical, electrical or mechanical. A biosensor

© The Author(s), under exclusive license to Springer Nature Switzerland AG 2021
E. Bhattacharya, *Biosensing with Silicon*, SpringerBriefs in Materials,
https://doi.org/10.1007/978-3-030-92714-1_1

consists of two parts: a bioreceptor module which is specific to the particular bioana-lyte under detection and a transducer module that converts the biological signals into, say electrical signals. The electrical signals can be further processed, to determine the concentration of the bioanalyte, by comparison with some reference calibration. Biosensors can be classified as electrochemical, optical, calorimetric or mass sensi-tive devices, according to the mechanism for the biological specificity or the mode of physio-chemical signal transduction. Suitable biological elements such as enzymes, receptors, immuno-chemicals, antibodies, etc. can be interfaced to any of these types of transducers to detect and estimate an appropriate analyte.

Some of the well-established analytical techniques available for biological and chemical detection are polymerized chain reaction (PCR), enzyme-linked immunoassay (ELISA), Fluorescence microscopy—to name a few. These techniques have been very successful for biological sensing and detection in a wide variety of fields such as medical diagnostics, environmental monitoring, chemical detection and for fundamental research. While these methods are reliable, there is a lacuna and hence the room for improvement in some areas like the requirement of (i) substantial quantities of samples and reagents (ii) requiring fluorescent labels, (iii) portability, (iv) expensive equipment that need (v) skilled operators and so on. Miniaturisation of the conventional diagnostic laboratory into labs-on-chip, with the help of the biosen-sors is an attractive possibility. Many of the conventional technologies have been miniaturized already like qPCR [6] and handheld Raman spectroscopy [7, 8], but the initial capital cost associated with them are still high. Typical key features for a successful biosensor would be: small size and portability, capability of label-free detection, flexibility to operate in different conditions and the ability to give accurate and precise results rapidly [9, 10].

1.2   Silicon Biosensors

Silicon, a work horse of microelectronics and Integrated Circuits, is also used in a wide variety of application-specific sensor systems [11] including the possibility of lab-on-a-chip. Though silicon is not inexpensive, due to its superior performance, biosensors made from silicon can be cost effective when mass produced. In addition, silicon biosensors can be miniaturised and used with sample volumes as low as a few microliters with the well-established silicon MicroElectroMechanical system (MEMS) technology. This provides a platform for developing highly sensitive, low cost and mass-produced miniaturised sensors and systems with a wide range of operation [12, 13]. Thus, silicon can mitigate many of the deficiencies in conven-tional diagnostics and the available technology can be used to develop handheld and portable devices for detection and sensing. Ease in integrating with control and ampli-fying ICs with the possibility of transmitting the collected data for further analysis makes these even more attractive from the point of care perspective.

1.3 Electrochemical Biosensors

Electrochemical biosensors are classified as (a) Amperometric—change in current, (b) Impedimetric—change in impedance, and (c) Potentiometric—change in potential. The principle of the amperometric method is, at a constant potential, the current flow between the working electrode and the reference electrode in a biochemical reaction is proportional to the concentration of the analyte under study. Many silicon biosensors work on the principle of change in impedance or surface charge when they come in contact with an electrolyte and thereby develop a potential across the solid–liquid interface and are classified as potentiometric biosensors. A potentiometric sensor consists of an ion selective working electrode and a reference/counter electrode placed in an analyte and the potential difference between the electrodes can be measured. It is easy to implement a potentiometric sensor in silicon in the form of a field effect device. Ion Sensitive Field Effect Transistors (ISFETs) and Electrolyte Insulator Semiconductor Capacitors (EISCAPs) are silicon field effect based potentiometric sensors used in the determination of bio-analytes like Triglycerides [14, 15] and Urea [16] through an enzymatic reaction. One of the biosensors discussed here is of the potentiometric type where the ion selective working electrode was implemented by forming a microreactor in silicon [17] and the reference/counter electrode was implemented by forming thin film electrode layer on glass [18].

1.4 Microcantilever Biosensors

Microcantilevers are simple MEMS structures, which provide a very effective platform for biosensing application because of their large surface area-to-volume ratio, allowing ultra high sensitivity for mass detection from measurements of resonance frequency [19, 20]. Mechanical structures as chemical sensors can be traced back to much before the advent of MEMS [21]. However, the simple visual readout often used for macroscale mechanical transducers in the early studies [22, 23] could not provide viable accuracy and sensitivity [24]. The macroscale cantilever transducers, due to their large, suspended masses and low resonance frequencies were also extremely susceptible to external vibrations. Hence, cantilever transducers had little practical appeal until both microscopic cantilevers and more precise means for their readout became widely available.

1.5   Triglyceride Sensors

Cardiovascular disease (CVD) is the major cause of human mortality in the world [25] with Asian countries leading in the globe [26]. In human blood lipid profile, Triglycerides (TGs) represent a biomarker of CVD [27, 28]. The blood lipid profile consists of LDL, HDL, TGs and total cholesterol. Monitoring of TGs in blood reduces many health risks and a TG level between 50 and 150 mg/dL is considered to be the normal clinical range. Table 1.1 shows the TG levels and their severity levels [29] (https://triglycerideslevels.org/).

TGs are esters composed of one glycerol molecule and three fatty acid molecules and can be enzymatically hydrolysed into these components. Conventionally, TGs are determined from the glycerol molecule concentration present using spectrophotometric or spectroflurometry techniques [30] (https://www.biovision.com/https:// www.biovision.com/). High performance liquid chromatography (HPLC) method was also used in the determination of TGs in blood serum [31]. The conventional methods for the determination of TGs are mostly laborious, time consuming and expensive.

Amperometric sensors often need extensive electrode modification and immobilization of multienzymes for coupled reactions [32, 33, 34]. Nanostructured conducting polymers have been used in impedimetric biosensors [35].

Most silicon based TG biosensors come under the classification of potentiometric sensors, Table 1.2 compares the performance of some of these.

More recently, several Amperometric TG sensors, with improved response time as well as detection limit, have been reported. The various stable working electrode used include Platinum (Pt), carbon (C), gold (Au) and Indium Tin Oxide (ITO), the suitability criterion being the ease of immobilisation on the electrode materials. Rosli et al., have used screen printed carbon electrode (SPCE) [41]. Rezvani et al. [42] reported use of graphite/AC/lipase/CHIT as working electrode, Pt as a counter electrode and Ag/AgCl as reference electrode [42]. For potentiometric sensors too, a reference electrode is required to measure the potential variation in zero current condition. Table 1.3 compares the performance of some recently reported TG sensors.

Table 1.1 Severity level for different TG levels [courtesy MS Veeramani]

Severity level	TG in mg/dL
Normal	<150
Borderline high	150–199
High	200–499
Very high	>500

Table 1.2 Comparison of analytical parameters of reported potentiometric TG biosensors [courtesy MS Veeramani]

Reference	Immobilization surface	Type of device	Linear range	Response time
Nakako et al. [36]	Insulated gate	ISFET	100 to 400 mM	2 min
Pijanowska et al. [15]	Nitrocellulose sheet and silica gel beads	ISFET	upto 30 mM for Triacetin and 4 mM for Tributyrin	<5 min
Vijayalakshmi et al. [37]	NiFe$_2$O$_4$ magnetic nanoparticles	ISFET	5 to 30 mM	15 min
Kumar Reddy et al. [38]	Oxidized porous Si	EISCAP	6 to 21 mM	15 min
Basu et al. [39]	Nitride surface	EISCAP	5 to 15 mM	30 min
Vemulachedu et al. [40]	Nitride surface	Miniaturized EISCAP	upto 5 mM	3 min 30 s

Table 1.3 Comparison of some recently reported TG biosensors [courtesy Abhishek Kumar]

Author	Type of sensor	Type of immobilization	Type of electrode	Detection limit	Response time (sec)
Rezvani et al. [42]	Amperometric	Graphite/AC/li pase/CHIT	Graphite	99 mg/L	
Solanki et al. [43]	Amperometric	ITO/chitosan-Z rO2/lipase	ITO	155 mg/L	45
Hasanah et al. [44]	Optical	Pectin hydrogel membrane with ETH5294 chromoionoph ore	Pectin/Cl	150 mg/L	300
Dhand et al. [45]	Impedimetric	ITO/PANI-NT/li pase	ITO		20
Rosli et al. [41]	Electrochemical	SPCE/f C/[C2 OC2 C1 pyrr][N(Tf)2]-GA-Lipase	SPCE	0.68 mM	46
Xu et al. [46]	Electrochemical/amperometric	GCE/(ε-polyly sine-heparin)- Lipase	GCE (Glassy Carbon Electrode)	0.67 mg dL-1	
Mondal et al. [47]	Amperometric	ITO/AgCNFs	ITO	10.6 mg/dL	10
Yücel et al. [33]	Amperometric	Gelatin membrane	GCE	120 mg/L	

1.6　Urea Sensor

Urea and ammonia are important analytes found in blood and other bodily fluids and are used as markers for detection of various infectious diseases besides renal and hepatic disorders in the body [48]. They are also active ingredients in pesticides which pose a threat of polluting the water bodies [49]. Therefore, detection of these analytes are important for applications such as diagnostic, food and environmental monitoring [50].

The urea biosensors currently in use employ electrochemical, optical and conductometric methods of detection. These sensors depend on the ammonium ion produced during the reaction for changes in conductivity/potential and hence face interference from the presence of ions like Na^+ and K^+. Also, for the amperometric sensors, an additional enzyme glutamate dehydrogenase is essential to cascade the current output [51]. The sensitivity and response time of potentiometric urea sensors depends upon the enzyme activity over the electrode surface and the rate of hydrolysis of the urea molecules. Thus the enzyme immobilisation techniques plays an important role. Das and Yoon [52] reported that urease enzyme-immobilized over sulfonated graphene/polyaniline nanocomposite film has high detection sensitivity $(0.85 \, \mu A \cdot cm^{-2} \cdot mM^{-1})$ and limit $(0.050 \, mM)$. More complex media have also been used to detect urea where, instead of using ammonium ions as the transducing factor, CO_2 gas was used to detect urea. Fapyane et al. [53] used a carbon dioxide microsensor, adjacent to the urease immobilised electrode in a glass chamber to estimate urea by sensing the CO_2 emission. Some of the recent urea biosensors are compared in Table 1.4.

Frequency-based methods for the detection of urea are limited in number. The commonly used devices for frequency-based detection includes quartz crystal microbalance, piezo-electric resonator, surface acoustic resonator etc. and these methods rely upon the measurement of change in pH or conductivity caused during the enzymatic reaction with urease.

In this monograph, the technology behind the device fabrication, the functionalization protocols leading to stable and reproducible immobilization, the measurement system and the data analysis for two specific silicon biosensors and bioMEMS are discussed. Many of the issues and challenges that crop up are not uncommon and can be extended to other family of sensors as well.

Table 1.4 Comparison of some recently reported urea biosensors [courtesy Abhishek Kumar]

Author	Type of sensor	Type of immobilization	Type of electrode	Detection limit	Sensitivity	Response time (sec)
Kim et al. [54]	Amperometric	SF/Glutaraldehyde/Urease	SF		4.62 mA/M·c m2	
Hsu et al. [55]	Potentiometric	Pt/PEDOT/BSA/urease	Pt		15.2 mV/deca de	10
Das and Yoon [52]	Amperometric	SGO/PANI/Ure ase	SGO (Sulfonated Graphene Oxide)	0.050 mM	0.85 μA cm $-$ 2·m M $-$ 1	5
Fapyane et al. [53]	Amperometric with CO_2 microsensor	Alginate Polymer/Urease	Alginate Polymer	1 μM	0.6 pA/μM	120
Lai et al. [56]	Potentiometric	ITO/P(3HT-co-3 TAA)/Urease	ITO	5 mM		360
Esmaeili et al. [57]	Potentiometric	Ag/AgCl/KC-KCl/Urease	Ag/AgCl treated with H + ionophore	0.001 mM	90.7 mV/ decad e	120
Öndeş et al. [58]	Potentiometric	SPCE/Poly(HE MA-GMA)NP/Ur ease	SPCE	0.77 μM		30
Rahmanian et al. [59]	Impedimetric	FTO/TiO2/Nano-ZnO/Urease	FTO (Fluorinated Tin Oxide)	2 mg dl $-$ 1	0.02 k per mg dL $-$ 1	Less than 4 s

References

1. Kawamura A, Miyata T (2016) Biosensors. Biomater Nanoarchitectonics 157–176. https://doi.org/10.1016/B978-0-323-37127-8.00010-8
2. Yoon J-Y (2013) Introduction to biosensors, 1st edn. Springer, New York
3. Narang J, Pundir CS (2017) Biosensors : an introductory textbook, 1st edn. Pan Stanford Publishing
4. Wang J (2001) Glucose biosensors: 40 years of advances and challenges. Electroanalysis 13:983–988. https://doi.org/10.1002/1521-4109(200108)13:12
5. Clark L, Lyons C (1962) Electrode systems for continuous monitoring in cardiovascular surgery. Ann N Y Acad Sci 102:29–45. https://doi.org/10.1111/J.1749-6632.1962.TB13623.X
6. Homola J (2008) Surface plasmon resonance sensors for detection of chemical and biological species. Chem Rev 108:462–493. https://doi.org/10.1021/CR068107D
7. Xin Jack Z, Daoguo J, Jun X et al (2013) Handheld Raman spectrometer
8. Zhou XJ, Jiang D, Xu J et al (2013) Handheld Raman Spectrometer, (US Patent -US D677,185S), pp 1–3
9. Hansen KM, Thundat T (2005) Microcantilever biosensors. Methods 37:57–64. https://doi.org/10.1016/J.YMETH.2005.05.011
10. Salehi-Khojin A, Bashash S, Jalili N et al (2009) Nanomechanical cantilever active probes for ultrasmall mass detection. J Appl Phys 105. https://doi.org/10.1063/1.3054371
11. Thevenot DR, Tóth K, Durst RA, Wilson GS (1999) Electrochemical biosensors: recommended definitions and classification. Pure Appl Chem 71:2333–2348. https://doi.org/10.1351/PAC199971122333
12. Tseytlin YM (2005) High resonant mass sensor evaluation: an effective method. Rev Sci Instrum 76. https://doi.org/10.1063/1.2115207
13. Carrascosa LG, Moreno M, Álvarez M, Lechuga LM (2006) Nanomechanical biosensors: a new sensing tool. TrAC Trends Anal Chem 25:196–206. https://doi.org/10.1016/J.TRAC.2005.09.006
14. Hierlemann A, Lange D, Hagleitner C et al (2000) Application-specific sensor systems based on CMOS chemical microsensors. Sens Actuat B Chem 70:2–11. https://doi.org/10.1016/S0925-4005(00)00546-3
15. Pijanowska DG, Baraniecka A, Wiater R et al (2001) The pH-detection of triglycerides. Sens Actuat B Chem 78:263–266. https://doi.org/10.1016/S0925-4005(01)00823-1
16. Pijanowska DG, Torbicz W (1997) pH-ISFET based urea biosensor. Sens Actuat B Chem 44:370–376. https://doi.org/10.1016/S0925-4005(97)00194-9
17. Vemulachedu H, Fernandez RE, Bhattacharya E, Chandha A (2008) Miniaturization of EISCAP sensor for triglyceride detection. J Mater Sci Mater Med 201(20):229–234. https://doi.org/10.1007/S10856-008-3534-Y
18. Veeramani MS, Shyam P, Ratchagar NP et al (2013) A miniaturized pH sensor with an embedded counter electrode and a readout circuit. IEEE Sens J 13:1941–1948. https://doi.org/10.1109/JSEN.2013.2245032
19. Craighead H (2006) Future lab-on-a-chip technologies for interrogating individual molecules. Nature 442:387–393. https://doi.org/10.1038/nature05061
20. Dittrich PS, Kaoru Tachikawa A, Manz A (2006) Micro total analysis systems. latest advancements and trends. Anal Chem 78:3887–3907. https://doi.org/10.1021/AC0605602
21. Timoshenko S (1930) Strength of materials; part i elementary theory and problems; part II advanced theory and problems, 1st edn. D. Van Nostrand Company Inc., New York, p 1930
22. Shaver PJ (1969) Bimetal strip hydrogen gas detectors. Rev Sci Instrum 40:901. https://doi.org/10.1063/1.1684100
23. Taylor EH, Waggener WC (1979) Measurement of adsorptive forces. J Phys Chem 83:1361–1362. https://doi.org/10.1021/J100473A025
24. Lavrik NV, Sepaniak MJ, Datskos PG (2004) Cantilever transducers as a platform for chemical and biological sensors. Rev Sci Instrum 75:2229–2253. https://doi.org/10.1063/1.1763252

25. W.H.O (2021) Cardiovascular Diseases. In: World Heal. Organ. https://www.who.int/newsroom/fact-sheets/detail/cardiovascular-diseases-(cvds)
26. Ueshima H, Sekikawa A, Miura K et al (2008) Cardiovascular disease and risk factors in asia: a selected review. Circulation 118:2702. https://doi.org/10.1161/CIRCULATIONAHA.108.790048
27. Hokanson JE, Austin MA (1996) Plasma triglyceride level is a risk factor for cardiovascular disease independent of high-density lipoprotein cholesterol level: a metaanalysis of population-based prospective studies. J Cardiovasc Risk 3:213–219. https://doi.org/10.1177/174182679600300214
28. Miller M, Stone NJ, Ballantyne C et al (2011) Triglycerides and cardiovascular disease. Circulation 123:2292–2333. https://doi.org/10.1161/CIR.0B013E3182160726
29. https://triglycerideslevels.org/ What Are Triglycerides Levels and Why Do They Matter? - Triglycerides Levels. https://triglycerideslevels.org/. Accessed 12 Sep 2021
30. https://www.biovision.com/ Triglyceride Quantification Colorimetric/Fluorometric Kit | K622 | BioVision, Inc. https://www.biovision.com/triglyceride-quantification-colorimetric-fluorometric-kit.html. Accessed 10 Sep 2021
31. Asmis R, Bühler E, Jelk J, Gey K (1997) Concurrent quantification of cellular cholesterol, cholesteryl esters and triglycerides in small biological samples. Reevaluation of thin layer chromatography using laser densitometry. J Chromatogr B Biomed Sci Appl 691:59–66. https://doi.org/10.1016/S0378-4347(96)00436-7
32. Solanki PR, Dhand C, Kaushik A et al (2009) Nanostructured cerium oxide film for triglyceride sensor. Sens Actuat B Chem 141:551–556. https://doi.org/10.1016/j.snb.2009.05.034
33. Yücel A, Özcan HM, Sağıroğlu A (2016) A new multienzyme-type biosensor for triglyceride determination. Prep Biochem Biotechnol 46:78–84. https://doi.org/10.1080/10826068.2014.985833
34. Pundir CS, Aggarwal V (2017) Amperometric triglyceride bionanosensor based on nanoparticles of lipase, glycerol kinase, glycerol-3-phosphate oxidase. Anal Biochem 517:56–63. https://doi.org/10.1016/j.ab.2016.11.013
35. Dhand C, Solanki PR, Datta M, Malhotra BD (2010) Polyaniline/single-walled carbon nanotubes composite based triglyceride biosensor. Electroanalysis 22:2683–2693. https://doi.org/10.1002/ELAN.201000269
36. Nakako M, Hanazato Y, Maeda M, Shiono S (1986) Neutral lipid enzyme electrode based on ion-sensitive field effect transistors. Anal Chim Acta 185:179–185. https://doi.org/10.1016/0003-2670(86)80044-7
37. Vijayalakshmi A, Tarunashree Y, Baruwati B et al (2008) Enzyme field effect transistor (ENFET) for estimation of triglycerides using magnetic nanoparticles. Biosens Bioelectron 23:1708–1714. https://doi.org/10.1016/J.BIOS.2008.02.003
38. Kumar Reddy RR, Basu I, Bhattacharya E, Chadha A (2003) Estimation of triglycerides by a porous silicon based potentiometric biosensor. Curr Appl Phys 3:155–161. https://doi.org/10.1016/S1567-1739(02)00194-3
39. Basu I, Subramanian RV, Mathew A et al (2005) Solid state potentiometric sensor for the estimation of tributyrin and urea. Sens Actuat B Chem 107:418–423. https://doi.org/10.1016/J.SNB.2004.10.038
40. Vemulachedu H, Fernandez RE, Bhattacharya E, Chadha A (2009) Miniaturization of EISCAP sensor for triglyceride detection. J Mater Sci Mater Med 20:229–234. https://doi.org/10.1007/s10856-008-3534-y
41. Rosli NH, Ahmad NM, Rani MAA et al (2017) The electroanalytical detection of triglyceride concentrations in olive oil using modified screen printed electrodes with ionic liquid [1-(2-ethoxyethyl)-1-methylpyrrolidinium bis(trifluoromethylsulfonyl)imide]-lipase. Int J Biosens Bioelectron 2:162–164. https://doi.org/10.15406/IJBSBE.2017.02.00040
42. Rezvani M, Najafpour GD, Mohammadi M, Zare H (2017) Amperometric biosensor for detection of triglyceride tributyrin based on zero point charge of activated carbon. Turkish J Biol 41:268–277. https://doi.org/10.3906/biy-1607-24

43. Solanki S, Pandey CM, Soni A et al (2015) An amperometric bienzymatic biosensor for the triglyceride tributyrin using an indium tin oxide electrode coated with electrophoretically deposited chitosan-wrapped nanozirconia. Microchim Acta 1831(183):167–176. https://doi.org/10.1007/S00604-015-1618-1

44. Hasanah U, Sani NDM, Heng LY et al (2019) Construction of a hydrogel pectin-based triglyceride optical biosensor with immobilized lipase enzymes. Biosensors 9:135. https://doi.org/10.3390/BIOS9040135

45. Dhand C, Solanki PR, Sood KN et al (2009) Polyaniline nanotubes for impedimetric triglyceride detection. Electrochem Commun 11:1482–1486. https://doi.org/10.1016/J.ELECOM.2009.05.034

46. Xu T, Chi B, Chu M et al (2018) Hemocompatible ε-polylysine-heparin microparticles: a platform for detecting triglycerides in whole blood. Biosens Bioelectron 99:571–577. https://doi.org/10.1016/J.BIOS.2017.08.030

47. Mondal K, Ali MA, Singh C et al (2017) Highly sensitive porous carbon and metal/carbon conducting nanofiber based enzymatic biosensors for triglyceride detection. Sens Actuat B Chem 246:202–214. https://doi.org/10.1016/J.SNB.2017.02.050

48. Fassett RG, Venuthurupalli SK, Gobe GC et al (2011) Biomarkers in chronic kidney disease: a review. Kidney Int 80:806–821. https://doi.org/10.1038/ki.2011.198

49. Belisle BS, Steffen MM, Pound HL et al (2016) Urea in lake erie: organic nutrient sources as potentially important drivers of phytoplankton biomass. J Great Lakes Res 42:599–607. https://doi.org/10.1016/j.jglr.2016.03.002

50. Testani JM, Cappola TP, Brensinger CM et al (2011) Interaction between loop diuretic-associated mortality and blood urea nitrogen concentration in chronic heart failure. J Am Coll Cardiol 58:375–382. https://doi.org/10.1016/j.jacc.2011.01.052

51. Dhawan G, Sumana G, Malhotra BD (2009) Recent developments in urea biosensors. Biochem Eng J 44:42–52. https://doi.org/10.1016/j.bej.2008.07.004

52. Das G, Yoon HH (2015) Amperometric urea biosensors based on sulfonated graphene/polyaniline nanocomposite. Int J Nanomedicine 10:55–66. https://doi.org/10.2147/IJN.S88315

53. Fapyane D, Berillo D, Marty J-L, Revsbech NP (2020) Urea biosensor based on a CO2 microsensor. ACS Omega 5:27582–27590. https://doi.org/10.1021/ACSOMEGA.0C04146

54. Kim K, Lee J, Moon BM et al (2018) Fabrication of a urea biosensor for real-time dynamic fluid measurement. Sensors 18:2607. https://doi.org/10.3390/S18082607

55. Hsu C-H, Hsu Y-W, Weng Y-C (2016) A novel potentiometric sensor based on urease/bovine serum albumin-poly(3,4-ethylenedioxythiophene)/Pt for urea detection. Zeitschrift Für Naturforsch B 71:277–282. https://doi.org/10.1515/ZNB-2015-0166

56. Lai CK, Foot PJ, Brown JW, Spearman P (2017) A Urea Potentiometric Biosensor Based on a Thiophene Copolymer. Biosensors 7:13. https://doi.org/10.3390/BIOS7010013

57. Esmaeili C, Heng LY, Ling YP et al (2017) Potentiometric Urea Biosensor Based on Immobilization of Urease in Kappa-Carrageenan Biopolymer. Sens Lett 15:851–857. https://doi.org/10.1166/SL.2017.3882

58. Öndeş B, Akpınar F, Uygun M et al (2021) High stability potentiometric urea biosensor based on enzyme attached nanoparticles. Microchem J 160. https://doi.org/10.1016/J.MICROC.2020.105667

59. Rahmanian R, Mozaffari SA, Amoli HS, Abedi M (2018) Development of sensitive impedimetric urea biosensor using DC sputtered Nano-ZnO on TiO2 thin film as a novel hierarchical nanostructure transducer. Sensors Actuators B Chem 256:760–774. https://doi.org/10.1016/J.SNB.2017.10.009

Chapter 2
Technology/Nuts and Bolts

Basic microelectronics processing technology as well as some micromachining processes for silicon, used for fabricating the biosensors, are briefly discussed. This includes unit processes like oxidation, doping, lithography, chemical vapour deposition and etching. Two standard MEMS process technologies, namely bulk and surface micromachining, are described. The required physical and electrical characterisation methods for biosensors are given. The protocol developed for the immobilization of the enzymes on the sensor surface is discussed.

There are many wonderful textbooks and references available on silicon processing for microelectronics [1–3] as well as for MEMS technology [4, 5]. Most silicon processes have to be carried out in temperature and humidity controlled Clean Rooms, characterized by the maximum number of particles/m^3. For eg. a Class 100 (ISO 5) clean room must have less than or equal to 100,000 particles/m^3. In this chapter, for the sake of completeness, we will briefly review the technology relevant to the biosensors and MEMS discussed here. We start with the basic unit process steps required for silicon micromachining. The full fabrication process, optimized for the sensors, will be discussed in the relevant chapters. Protocols for surface functionalization, in this case immobilization of enzymes on the sensor surface, are also discussed.

2.1 Unit Processes

Wafer cleaning is the most important first step and is done following the standard RCA protocol [6–8]. This is used to remove organics, heavy metals and alkali ions and the native oxide from the surface of the silicon wafer.

© The Author(s), under exclusive license to Springer Nature Switzerland AG 2021
E. Bhattacharya, *Biosensing with Silicon*, SpringerBriefs in Materials,
https://doi.org/10.1007/978-3-030-92714-1_2

2.1.1 Oxidation

A major reason for the success of silicon in ICs is the ease in forming a stable oxide with an excellent interface. The many applications of silicon oxide are as an isolation layer between devices, for surface passivation, as a masking layer (for etching, etc.), and as the gate oxide in MOSFETs. The main two oxidation processes used are (i) thermal, where the oxide is grown by exposing the Si to oxygen at high temperatures (800–1000 °C) and (ii) chemical vapour deposition (CVD), where a relatively low temperature oxide (LTO) (450 °C) is deposited from gas phase precursors on the surface of the silicon wafer. Thermal oxidation, carried out in furnaces with quartz tubes, can be a wet process with water vapour as the source of oxygen or dry using oxygen gas. The slower growth rate dry oxide, with its high material density and breakdown voltage, is of excellent quality and is used as the gate oxide in MOSFETs with low leakage current. Wet oxides, on the other hand, have faster growth rates, lower density, dielectric strength and breakdown voltage. The quality of wet oxides is not as good as dry oxides but is better than the deposited oxides and they are commonly used for masking, isolation (called field oxide or FOX) and surface passivation. The oxidation rate depends on the properties of the wafer like the orientation and doping, with typical oxidation rates around 0.5 to 1 µm/hour.

Since thermal oxidation is a high temperature process of long duration, there can be significant dopant redistribution in the wafer due to diffusion during oxidation. An advanced oxidation technique to avoid this is using rapid thermal processing (RTP). In RTP, the wafer is rapidly heated from a low to a high processing temperature (T > 900 °C), held at that temperature for a short time and then brought back rapidly to a low temperature. Typical temperature ramp rates range from 10 to 350 °C/s, compared to about 0.1 °C/s for furnace processing. Since the duration of RTP at high temperatures is short, typically from 1 s to 5 min, it is also suitable to grow thin (<40 nm) gate oxides of high quality.

2.1.2 Doping

Changing the conductivity of silicon by introducing controlled amount of dopant atoms can be achieved by diffusion as well as ion implantation. Ion implantation, a versatile technique with the possibility to achieve a variety of doping profiles, requires expensive equipment and we have restricted ourselves to diffusion for doping. Diffusion is a high temperature process (around 1000 °C) driven by the existence of a concentration gradient and yields a doping profile with the highest concentration at the surface. The penetration depth depends on the diffusion length of the dopant and the exact profile is a Gaussian for a limited source process or a complementary error function for an infinite source process. Diffusion is aided by the presence of point defects and therefore is faster in poly silicon with its grain boundaries when compared to crystalline silicon.

2.1.3 Lithography

Lithography is a method to transfer patterns onto the surface of the wafer by exposing photosensitive polymers, called photoresist, through patterned masks followed by developing. UV light is used for photolithography and, at the μm scale, can be used to create rapidly and reproducibly structures of desired shape. For smaller feature size, e-beam or X-ray lithography can be used.

The basic steps in photolithography are: Application of photoresist, on the cleaned wafer surface, using a spinner—the speed of the spinner decides the thickness of the photoresist layer. After a soft bake, to remove the moisture and promote adhesion, the appropriate mask is aligned on the wafer and exposed to UV. Developing the photoresist, after the UV exposure, removes the soft photoresist and delineates the pattern. A hard bake is done to densify the remaining photoresist so it can withstand the subsequent step like etching of the exposed area. And finally, the remaining photoresist is stripped followed by a final inspection to remove any unwanted stringers left over from the photoresist.

2.1.4 Chemical Vapour Deposition (CVD) Processes

Chemical Vapour Deposition is the formation of a thin film on a substrate by the reaction of vapour phase chemicals or gases containing the required constituents. The reactant gases are introduced into a reaction chamber and they decompose and react on a heated surface to form the thin film. Low pressure CVD (LPCVD) is used routinely to deposit polycrystalline silicon (polySi) by thermally dissociating silane (SiH_4) or dichlorosilane (SiH_2Cl_2); silicon oxide (LTO referred to above) by reacting silane with oxygen; silicon nitride by reacting silane with ammonia. CVD films have good conformality or step coverage and typical deposition temperatures vary from 450 to 650 °C. Being thin films, the properties are strongly dependent on the deposition conditions and must be optimised for specific applications.

Plasma enhanced CVD (PECVD) is another technique where the deposition temperatures are lower than LPCVD. Here, the energy required for the reaction is provided partially by striking a plasma in the gas using an RF source. With an additional power source to confine the plasma, for example an inductively coupled plasma CVD, the deposition temperature can be lowered further. Films grown by LPCVD usually are superior to those grown by PECVD in terms of their density and electronic and mechanical properties.

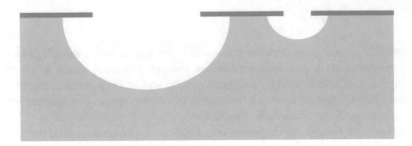

Fig. 2.1 Isotropic etching (courtesy A. Chakrabarti)

2.1.5 Etching Wet/Dry

Etching is an important process step in micromachining. Both wet and dry etching processes are used. In wet etching, as the name suggests, liquid etchants like acids or alkalis are used and the etching can be done in fume hoods using standard inexpensive laboratory apparatus like beakers and petri dishes. Care should be taken to dispose the corrosive waste products. In contrast, dry etching uses safer non-toxic gases like O_2, CF_4 and SF_6 in small quantities and is carried out in a plasma reactor where the waste products can be easily discharged. The equipment required are expensive but provides ease of automation.

Etching can be isotropic, where the etch rate is equal in all directions, as shown in Fig. 2.1, or anisotropic. Selectivity of the etchant, which is the ability to etch the desired layer but not etch the underlying substrate or the protective mask layer, is an important criterion satisfied in most chemical etch processes. The aspect ratio, which is the ratio of the etch geometries—depth to width—is another important parameter, especially for deep etching required for MEMS.

2.1.6 Silicon Wet Etching

The etching process involves three steps: (a) Transport of the reactants to the silicon surface, (b) Surface reaction and (c) Transport of the reaction products away from the surface. The most common etchant for silicon is HNA, which is a mixture of hydrofluoric acid, nitric acid and acetic acid. It is an isotropic etchant with a high etch rate of 50 μm/min even at room temperature, making it possible to use photoresist masks.

Though most wet chemical etch processes are isotropic, an important exception is etching of crystalline silicon wafers in bases like KOH. The anisotropy is because the etch rate in the (111) direction is much lower than in the (100) direction. It is believed that this is a consequence of the Si atoms on the (100) surface having two back-bonded Si atoms while the ones on the (111) surface have three [9]. Thus, the cross-section of a (100) silicon wafer etched in KOH through a window looks like Fig. 2.2, where the angle between the (100) and the (111) planes is 54.74°. The angle between the (110) and the (111) planes happens to be 90° and therefore it is easy to make structures with vertical sidewalls by simple wet etching, though (110) silicon wafers are more expensive.

Other than KOH, Ethylene Diamine Pyrocatechol (EDP) and Tetramethyl ammonium hydroxide (TMAH) are also anisotropic etchants for silicon. Usually, because of low etch rates (<1 μm)/min, temperatures between 80–100 °C is required for wet anisotropic etching and photoresist mask cannot be used. Typical masking films used are silicon nitride and oxide. Also, for reproducibility, care must be taken to ensure that the concentration of the etchant does not change due to evaporation at these temperatures—especially because the etching durations are long. The etch rate drops drastically when the doping concentration for p-type Si exceeds 10^{19} cm^{-3} [10].

SiO$_2$ wet etch can be done in HF with or without the addition of ammonium fluoride (NH$_4$F). Addition of the ammonium fluoride creates a buffered HF solution (BHF) also called as the buffered oxide etchant (BOE). Silicon Nitride can also be etched in BOE (7:1) but the etch rate is low at 200 nm/min. Hot phosphoric acid can etch nitride with a thin patterned oxide used as the etch mask. The etch rate at 165 °C is 550 nm/min and the selectivity is good with the etch rate for silicon dioxide at 100 nm/min and of silicon at 0.1 nm/min.

Dry etching is usually carried out in a plasma reactor where the reactant gases are let in and highly reactive free radicals are generated in the plasma for etching. The etch products are volatile and can be pumped out. Silicon can be etched by any halogen atom (Cl, Br, F, I). The non-toxic fluorine, using gases like CF$_4$ and SF$_6$, is typically preferred because it reacts with silicon spontaneously and the etching is isotropic. Anisotropic etching can be achieved either by ionic bombardment to damage the exposed surface, as in reactive ion etching (RIE), or the passivation of the sidewalls by an inhibitor as shown in Fig. 2.3a and b.

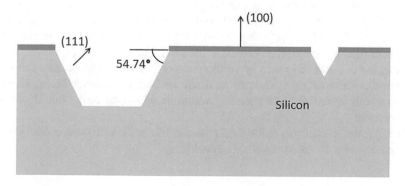

Fig. 2.2 Anisotropic etching of (100) Si wafer (courtesy A. Chakrabarti)

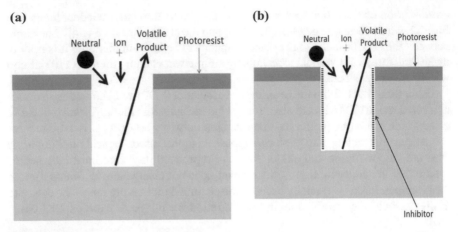

Fig. 2.3 a Energetic ion-enhanced etching and **b** Inhibitor ion-enhanced etching (adapted from Oxford Instruments brochure, courtesy A. Chakrabarti)

2.2 Micromachining

Two major processes to fabricate MEMS structures are bulk and surface micromachining. Though surface micromachining is often cited to be more IC compatible, and hence easier to integrate with measurement and control circuits, there are multiple instances of circuits integrated with bulk micromachined sensors too (Motorola pressure sensor). Each has its unique benefits and many sensors and actuators have been fabricated using a combination of the two processes.

Bulk micromachining is a subtractive process involving deep etching, sometimes from both sides of the wafer, to fabricate the desired structures. It typically has large features of substantial mass and thickness (in 100 s of μm) often involving use of multiple wafers. The important unit steps are deep etching and bonding of wafers, like silicon to silicon or silicon to glass. The miniaturized EISCAPs are fabricated using bulk micromachining of silicon and glass wafers followed by bonding.

Surface micromachining is an additive process where thin films of few μm thickness, acting as structural or sacrificial layers, are deposited on a substrate and the structures are formed by patterning and etching of those layers [11]. We use polySi as the structural layer and silicon oxide as the sacrificial layer. The cantilever sensors of thickness 2–3 μm are made by surface micromachining. Figure 2.4a shows the one mask process steps to fabricate an oxide anchored cantilever beam. The final release step is critical for free standing structures, Fig. 2.4b shows the SEM of the released beam.

The thinner cantilevers with thickness in 100 s of nm were made using a combination of bulk and surface micromachining.

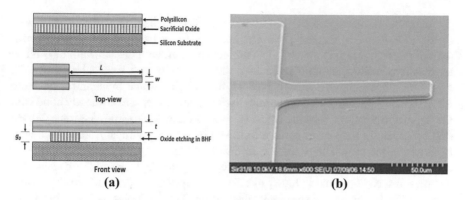

Fig. 2.4 **a** Surface micromachined cantilevers, **b** SEM of released beam (courtesy A. Chakrabarti and S. Bhat)

2.3 Characterisation

Physical characterization of the layers and structures were done using Ellipsometry, SEM and the surface profiler.

For electrical characterization, a probe station was used. Pull-in measurements on cantilevers were made using a dc power supply and a current meter. Capacitance (C-V) measurements were made using a bridge.

Resonance frequency of cantilevers was measured using a Laser Doppler vibrometer (LDV). A custom equipment called Cantisens® has also been used to measure resonance frequency, especially in liquids.

2.4 Surface Functionalization

Covalent immobilsation of enzymes

There are three major methods of surface immobilisation of biomolecules and cells [12]: (1) physical adsorption that uses Van der Waal's/electrostatic force, affinity, or adsorbed and crosslinked; (2) physical entrapment that uses a barrier or matrix systems; and (3) covalent attachment to the support molecules using solid surface chemistry [13]. A chemically immobilized biomolecule may be attached via a spacer group that can provide less steric hindrance and greater freedom of movement to the immobilized biomolecules leading to greater specific activity for the biomolecules. Enzyme immobilization by covalent binding has many advantages viz. tighter binding during the reaction and systems with immobilized enzymes are reported to have more heat stability [14]. Silane reactions, due to their simplicity and stability, are quite popular and can be used to modify hydroxylated or amine-rich surfaces.

An immobilisation protocol for the enzyme lipase on the silicon nitride was developed and is shown in Fig. 2.5 [15]. The enzyme (lipase) with amine (–NH$_2$) terminations can be covalently attached to a molecular layer, with the same amine (–NH$_2$) terminations, using glutaraldehyde as a linker molecule. A molecular layer with amine terminations (–NH$_2$) is formed on the sensor surface by silanisation using APTES. The linker molecule (glutaraldehyde), in principle, can covalently bond with the amine terminations forming imine (C = N) linkages. Enzymes are immobilised first by silanising the sensor surface followed by treating with activated glutaraldehyde and lipase [16]. The process is initiated by activating the nitride surface with conc. HNO$_3$ for 30 min at 80 °C. This step was later replaced with exposing the samples to a N$_2$O plasma [17, 18] in a PECVD system. This results in the formation of –OH bonds on the surface of the nitride. The N$_2$O plasma treatment helps to promote uniform covalent adhesion of APTES and other subsequent functional layers to the nitride surface of the sensor [19].

The functionalized surfaces were characterised using FTIR at each stage confirming the formation of hydroxyl, amine and imine bonds after each of the immobilisation process steps. The morphology of the enzyme immobilized surface was characterized using scanning electron microscopy (SEM). The activity of the enzyme P. cepacia lipase was determined by the pNPB assay [15, 20]. The specific activities of the free and immobilised enzyme kinetics were found to be 37.8 and 28.95 U/mg (one enzyme unit 'U' defined as the amount of enzyme that converts 1 mM of pNPB into product per minute). The immobilised enzyme retained 76.58% of its native activity. A similar protocol was followed for the immobilisation of the enzyme urease [21].

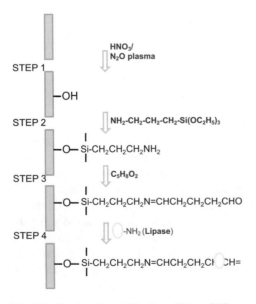

Step1 : Samples are treated with HNO$_3$ or N$_2$O plasma

Step 2: APTES is used for silanisation

Step 3: Glutaraldehyde treatment

Step 4 : Enzyme (Lipase) immobilization

Fig. 2.5 Covalent immobilization of lipase [15]

References

1. Ghandhi SK (1994) VLSI fabrication principles: silicon and gallium arsenide, 2nd edn. Wiley
2. Plummer JD, Deal M, Griffen PB (2000) Silicon VLSI technology: fundamentals, practice and modeling. Pearson Education
3. Sze S (1988) VLSI technology. McGraw-Hill
4. Madou MJ (2011) Fundamentals of microfabrication and nanotechnology. CRC Press
5. Senturia SD (2001) Microsystem design. Springer
6. Kern W (1970) Radiochemical study of semiconductor surface contamination. I. Adsorpt Reag Compon RCA Rev 31:207–233
7. Kern W (1990) The evolution of silicon wafer cleaning technology. Proc—Electrochem Soc 137:1887–1892. https://doi.org/10.1149/1.2086825
8. Kern W, Puotinen DA (1970) Cleaning solutions based on hydrogen peroxide for use in silicon semiconductor technology. RCA Rev 31:187–206
9. Seidel H, Csepregi L, Heuberger A, Baumgärtel H (1990) Anisotropic etching of crystalline silicon in alkaline solutions: I. orientation dependence and behavior of passivation layers. J Electrochem Soc 137:3612–3626. https://doi.org/10.1149/1.2086277
10. Seidel H, Csepregi L, Heuberger A, Baumgärtel H (1990) Anisotropic etching of crystalline silicon in alkaline solutions: II. Infl Dopants J Electrochem Soc 137:3626–3632. https://doi.org/10.1149/1.2086278
11. Howe RT (1988) Surface micromachining for microsensors and microactuators. J Vac Sci Technol B Microelectron Nanom Struct 6:1809. https://doi.org/10.1116/1.584158
12. Davis DH, Giannoulis CS, Johnson RW, Desai TA (2002) Immobilization of RGD to ⟨111⟩ silicon surfaces for enhanced cell adhesion and proliferation. Biomaterials 23:4019–4027. https://doi.org/10.1016/S0142-9612(02)00152-7
13. Hoffman AS (1992) Immobilization of biomolecules and cells on and within polymeric biomaterials. Biologically modified polymeric biomaterial surfaces. Springer, Netherlands, pp 61–65
14. Varavinit S, Chaokasem N, Shobsngob S (2001) Covalent immobilization of a glucoamylase to bagasse dialdehyde cellulose. World J Microbiol Biotechnol 17:721–725. https://doi.org/10.1023/A:1012984802624
15. Fernandez RE, Bhattacharya E, Chadha A (2008) Covalent immobilization of Pseudomonas cepacia lipase on semiconducting materials. Appl Surf Sci 254:4512–4519. https://doi.org/10.1016/j.apsusc.2008.01.099
16. Howarter JA, Youngblood JP (2006) Optimization of silica silanization by 3-aminopropyltriethoxysilane. Langmuir 22:11142–11147. https://doi.org/10.1021/la061240g
17. Bose M, Basa DK, Bose DN (2000) Study of nitrous oxide plasma oxidation of silicon nitride thin films. Appl Surf Sci 158:275–280. https://doi.org/10.1016/S0169-4332(00)00023-4
18. Veeramani MS, Shyam KP, Ratchagar NP et al (2014) Miniaturised silicon biosensors for the detection of triglyceride in blood serum. Anal Methods 6:1728–1735. https://doi.org/10.1039/c3ay42274g
19. Te YL, Chou JC, Chung WY et al (2001) Characteristics of silicon nitride after O$_{2}$ plasma surface treatment for pH-ISFET applications. IEEE Trans Biomed Eng 48:340–344. https://doi.org/10.1109/10.914797
20. Shirai K, Jackson RL (1982) Lipoprotein lipase-catalyzed hydrolysis of p-nitrophenyl butyrate. Interfacial activation by phospholipid vesicles. J Biol Chem 257:1253–1258
21. Fernandez RE, Bhattacharya E, Chadha A (2010) Dynamic response of polysilicon microcantilevers to enzymatic hydrolysis of urea. Int J Adv Eng Sci Appl Math 2:17–22. https://doi.org/10.1007/s12572-010-0007-6

Chapter 3
Electrolyte Insulator Semiconductor Capacitor (EISCAP) Biosensor

Electrolyte Insulator Semiconductor capacitors (EISCAPs) are sensitive to changes in the pH of the electrolyte which can be sensed as shifts in the capacitance–voltage characteristics of the device. The sensitivity to pH can be utilized for biosensing as the enzymatic hydrolysis of many bioanalytes produce acids and bases, changing the pH. The EISCAP is optimized for the detection of triglycerides with the enzyme lipase and urea with the enzyme urease. Good agreement was found between the measurements with the EISCAP and clinical measurements. The EISCAP was miniaturized using micromachining techniques. With the enzyme immobilized on the Insulator surface, a compact biosensor chip was fabricated and characterized.

The Electrolyte Insulator Semiconductor capacitor (EISCAP) is an electrochemical potentiometric sensor. It is a simple two terminal device that is sensitive to changes in the pH of the electrolyte. The EISCAP can be thought of as the familiar and well understood Metal Oxide Semiconductor (MOS) capacitor, with the metal layer replaced by the electrolyte. An ion sensitive field effect transistor (ISFET) with the electrolyte as the gate was first introduced by [1]. A typical nMOS capacitor on a p-type silicon substrate is shown in Fig. 3.1a and b shows an EISCAP. Siu and Cobbold [2] explained the quasi equilibrium characteristics of the EIS system. Just like the capacitance–voltage (CV) characteristics of the MOS capacitor shifts with the work function of the metal, the CV characteristics of the EISCAP shifts with the pH of the electrolyte and this forms the basis of several such sensors as described in a recent review [3]. Many biological reactions are accompanied by the production of acids or bases, resulting in a change in the pH. Therefore, functionalising the dielectric surface with an appropriate biomolecule and tying the pH change with a specific biochemical reaction can transform the EISCAP into a biosensor that is both sensitive as well as selective. Enzymes, which act as catalysts, in the hydrolysis of many bioanalytes can serve this purpose—converting a general pH sensor to a specific biosensor.

© The Author(s), under exclusive license to Springer Nature Switzerland AG 2021
E. Bhattacharya, *Biosensing with Silicon*, SpringerBriefs in Materials,
https://doi.org/10.1007/978-3-030-92714-1_3

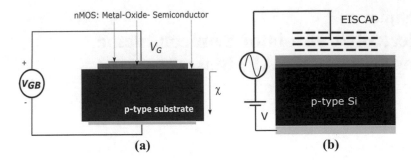

Fig. 3.1 An nMOS (**a**) and an EISCAP (**b**) on p-type silicon. The electrolyte in (**b**) replaces the metal in (**a**) (courtesy V.T. Fidal)

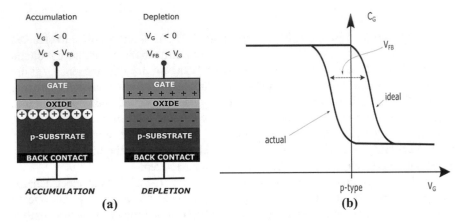

Fig. 3.2 **a** Ideal nMOS capacitor in accumulation and depletion modes ($V_{FB} = 0$) and **b** the corresponding high frequency CV characteristics in ideal and actual case when V_{FB} is not zero

Figure 3.2a shows the biasing of the nMOS and Fig. 3.2b shows the CV characteristics for the same, measured at high frequency with a small ac amplitude and varying DC bias. For the ideal nMOS of Fig. 3.1a, at zero applied bias V_G, the metal work function is such that the Fermi levels in the metal and the Si are aligned and there is no band bending, or the energy bands are flat in the semiconductor and we say that the flat-band voltage (V_{FB}) is zero. When $V_G < 0$ and the metal has negative charge, the free holes in the p-type Si accumulate at the oxide-Si interface and the resulting capacitance C_G, called the accumulation capacitance, is constant with a magnitude equal to the oxide capacitance. When $V_G > 0$ and the metal has positive charge, the free holes in the Si are repelled from the oxide-Si interface leaving a depletion region with negatively charged fixed acceptor ions. The total capacitance C_G is now a series combination of the oxide and the depletion layer capacitances and reduces. As V_G becomes more positive, to account for the increased negative charge, the depletion region becomes wider and the total capacitance value keeps decreasing as shown in Fig. 3.2b. In practice, the energy bands in Si at zero V_G may not be flat due to the metal work function value, the presence of oxide charge, interface charge, etc. and hence the bands are bent in Si even at $V_G = 0$. In this situation, a voltage V_{FB} has to be applied to get the flat band condition in silicon. The effect of this on the CV characteristics is shown in Fig. 3.2b where the CV shows a parallel shift long the voltage axis with the magnitude V_{FB}.

In a potentiometric electrochemical biosensor, a significant potential develops at the electrode surface due to the accumulation of charge. A dominant category of the potentiometric sensors is the ion selective electrodes (ISEs) based on thin films or selective membranes as recognition elements. Significant advantages of such sensors are the simplicity, possibility of miniaturization and mass production. Ion selective electrodes fall in two categories: the **I**on—**S**ensitive—**F**ield—**E**ffect—**T**ransistor (ISFET) and the **E**lectrolyte—**I**nsulator—**S**emiconductor—**CAP**acitor (EISCAP). ISFETs, though highly sensitive, can have problems such as poor adhesion and fast leaching-out of the sensitive materials, insufficient electrochemical stability due to corrosion of contacts requiring passivating layers to protect the electronic circuitry [4]. In contrast, EISCAPs, with their simpler fabrication process are more robust, exhibiting higher long-term stability. In EISCAPs, the insulator is in direct contact with the electrolyte solution. Depending upon the composition of the solution, an electric potential develops at the interface between the insulator and the solution.

3.1   Principle of Operation

Figure 3.3 shows the schematic of an EISCAP showing the insulator layer in direct contact with the electrolyte solution contained in a Teflon cell. The surface of the insulator has ionizable sites that directly interact with the electrolyte to either bind or release hydrogen ions [5].

This can be explained more clearly by considering the specific example of SiO_2. When SiO_2 is brought in contact with an aqueous solution, the surface of SiO_2 is

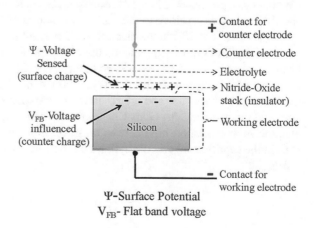

Fig. 3.3 Schematic of the EISCAP sensor [courtesy MS Veeramani]

Ψ -Voltage Sensed (surface charge)

V_{FB}-Voltage influenced (counter charge)

Silicon

Contact for counter electrode

Counter electrode

Electrolyte

Nitride-Oxide stack (insulator)

Working electrode

Contact for working electrode

Ψ-Surface Potential
V_{FB}- Flat band voltage

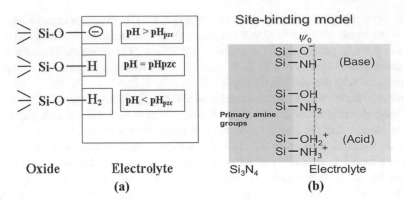

Fig. 3.4 Site binding model at the insulator/electrolyte interface of the EISCAP with **a** oxide and **b** nitride as insulators [courtesy MS Veeramani]

hydrolysed to form silanol (Si–OH) groups. For the dielectric-electrolyte interface, a point of zero charge (pH_{pzc}) is defined as that pH at which the dielectric surface has zero charge. Depending on the pH of the solution, the silanol groups on the dielectric surface can be positively charged for $pH < pH_{pzc}$, negatively charged for $pH > pH_{pzc}$ or neutral when $pH = pH_{pzc}$ as shown in Fig. 3.4a. The reaction creates a charge on the insulator surface that is compensated by an ionic counter charge in the electrolyte. This forms an electrical double layer in which a pH dependent potential drop occurs.

Thermal oxide on silicon can be grown with good control and has an excellent interface with the silicon substrate. But SiO_2, being inherently porous in nature, can let water molecules from the electrolyte permeate and reach the Si/oxide interface, as can the ions in the electrolyte. This would change the CV characteristics over time and therefore is not a reliable dielectric. Moreover, the silicon oxide layer can also get etched in basic solutions. Silicon nitride deposited by chemical vapour deposition (CVD), on the other hand, not only is more resistant to electrolyte spoiling but also exhibits higher pH sensitivity compared to the oxide. In this case, the primary amine (Si-NH$_2$) groups, shown in Fig. 3.4b, are the surface sites available on the nitride layer and they are more sensitive to the association and dissociation of H$^+$ ions than any other ions (Na$^+$ and K$^+$) in the electrolyte. The site-binding theory [5–8] originally proposed by Yates et al. [9] explains the presence of (Si-NH$_2$) sites on the surface of the insulator. Gouy Chapman-Stern theory [10–12] explains the formation of the electrical double layer at the insulator- electrolyte interface due to the availability of the insulator surface sites and the H$^+$ ion concentration in the electrolyte. From the site-binding theory and the Gouy Chapman- Stern theory, the Nernst response, or the electrochemical relationship between the surface potential (leading to the flat band shift in the capacitance) and the hydrogen ion concentration is given in Eqs. 3.1 and 3.2 as [2]:

$$\frac{\partial \psi}{\partial pH} = -2.3\frac{kT}{q}\alpha \tag{3.1}$$

Fig. 3.5 H⁺ ions bonding to surface sites on the insulator to influence the flat band voltage of the EISCAP [courtesy MS Veeramani]

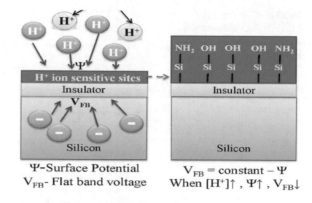

Ψ–Surface Potential
V_{FB}– Flat band voltage

V_{FB} = constant − Ψ
When [H⁺]↑ , Ψ↑ , V_{FB}↓

$$\text{Where } \alpha = \frac{1}{1 + \frac{\text{constant}}{\beta_{\text{int}}}} \tag{3.2}$$

The parameter β_{int} in Eq. (3.2) is the measure of the ability of the sensing layer (nitride) to absorb protons, in relation with the proton concentration at the sensing interface. As βint reaches infinity, α reaches unity and the change in surface potential (or flat band voltage) with change in pH at 25 °C reaches 59.2 mV/pH, the ideal pH sensitivity. The actual pH sensitivity of the EISCAP would be less than 59.2 mV/pH depending upon the ability of the nitride layer to sense proton concentration in the electrolyte. Figure 3.5 shows the insulator surface sites sensing the H⁺ ions and influencing the flat band voltage of the EISCAP.

The pH dependent flat band voltage of the EISCAP is given in Eq. 3.3 in terms of E_{Ref}, the electrode potential; ψ_o, the potential drop at the insulator/electrolyte interface; χ^{sol}, the surface dipole potential of the solution and φ_{Si}, the silicon work function. The increase in H⁺ ion concentration increases the surface potential (ψ_o) and the flat band voltage (V_{FB}) becomes more negative.

$$V_{FB} = E_{Ref} - \psi_0(pH) + \chi^{sol} - \frac{\phi_i}{q} \tag{3.3}$$

Hence, while characterizing the EISCAPs through C-V measurements with electrolytes of different pH, a pH dependent shift in C-V curves is observed as shown in Fig. 3.6. To quantify the shift, a point of constant capacitance is chosen at the midpoint of the depletion region of the CV curve and the voltage value at that point is termed as U_{bias}. The pH sensitivity is determined by plotting U_{bias} as a function of the pH.

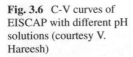

Fig. 3.6 C-V curves of EISCAP with different pH solutions (courtesy V. Hareesh)

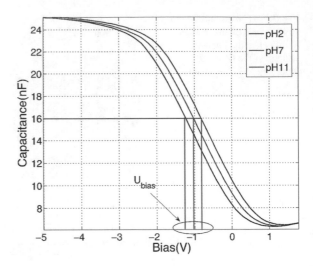

3.2 Fabrication of Sensors

The processing steps involved in the fabrication of EISCAP sensors on crystalline silicon substrates with stacked oxide-nitride dielectric are described [13]. 4–11 Ω-cm p-type (100) silicon substrates were oxidized at 1000 °C in dry oxygen for 2 h, followed by annealing at the same temperature in nitrogen ambient for 30 min. The oxide thickness measured by an ellipsometer was 100 nm. To get the stacked oxide-nitride structure, 80 nm silicon nitride was deposited over the oxidized silicon samples by plasma enhanced chemical vapour deposition (PECVD). The samples were then subjected to a 20 min anneal in nitrogen ambient at 800 °C [14]. The annealing step is essential because untreated silicon nitride, when kept exposed to ambient conditions, results in the oxidation of silicon nitride surface and degradation of the pH sensitivity of silicon nitride [2, 4]. To get an ohmic back contact, protecting the front side with photoresist, the back oxide was etched in buffered hydrofluoric acid (BHF) followed by the deposition of aluminium (Al) by thermal evaporation. For our initial work, this device together with the electrolyte containing the bioanalyte and the enzyme formed the biosensor.

3.3 Calibration and Measurements

A column of the electrolyte is contained on the top of the device with the help of a Teflon cell and O-ring as shown in Fig. 3.7a—a photograph is shown in Fig. 3.7b. The electrolyte is prepared in a buffer (0.01 M KH_2PO_4/ Na_2HPO_4) solution with KCl (1 M) as the ionic strength adjuster. A Platinum wire dipped in the electrolyte from the top was one contact while the other contact to the back Aluminium was taken

(a)

(b)

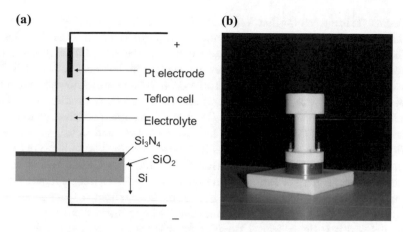

Fig. 3.7 **a** Schematic and **b** photograph of the EISCAP sensor [13]

through the bottom steel plate. C-V measurements were done at room temperature using the HP 4274A multifrequency LCR meter by sweeping the dc bias from a negative to a positive voltage, the amplitude and frequency of the ac signal being maintained constant at 15 mV and 4 kHz. The measurement frequency is important as a strong frequency dispersion is seen in the CV plots (Fig. 3.8), with the swing in the depletion region reducing at higher frequencies, which can be attributed to the effect of series resistance [15].

In MOSCAPs this kind of frequency dispersion is caused by the series resistance of the bulk silicon wafer [16]. However, for the EISCAPs, it is the series resistance of

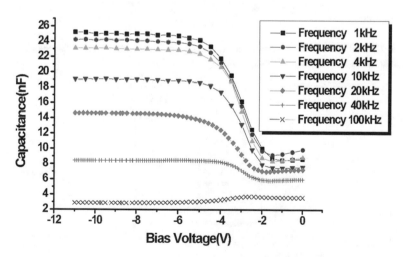

Fig. 3.8 Capacitance—Voltage curves at different frequencies with 1 M KCl electrolyte [courtesy I Basu]

the electrolyte in the column that dominates. This is verified by making measurements at 1 kHz and 10 kHz for varying KCl concentrations of 1, 2.5 and 5 M (Fig. 3.9).

The dispersion is less at lower frequencies, and the CV plots merge for KCl concentrations of 1–3 M. One can argue for using a high concentration of KCl to ensure that the series resistance is not a problem, but K^+ ions are a known contaminant for Si devices and hence it is advisable to use the lowest concentration of KCl that can still give a significant swing in the CV plot. To maintain constant conductivity of the electrolytes of different pH, all solutions are prepared in 0.1 M phosphate buffer with 0.5 M or 1 M KCl added as ionic strength adjuster with the pH values varying from 3.5 to 10 by mixing measured quantities of HCl and NaOH and the measured CV characteristics are given in Fig. 3.10a. Choosing a capacitance value in the middle of the depletion region (say at 70% of the maximum capacitance, C_{max}), the corresponding voltage U_{bias} was determined. Figure 3.10b shows a plot of U_{bias} vs the pH and it is linear with a slope of 55 mV for unit change in pH. Though the Nernst response predicts a sensitivity of 59 mV for unit change in pH, measured

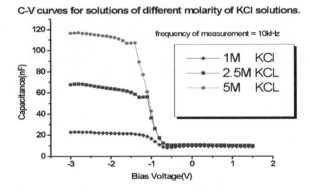

Fig. 3.9 Capacitance—Voltage curves with varying conductivity (molarity) of the electrolyte KCl at the same pH and frequency [courtesy I Basu]

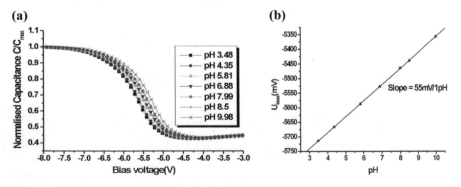

Fig. 3.10 a Capacitance –Voltage curves with varying pH of the electrolyte and **b** corresponding pH sensitivity [13]

values can be much lower. Harame et al. [5] suggested this could be because of partial oxidation of the nitride surface.

On an oxygen free nitride surface, each nitrogen atom can be assumed to be bonded to three Si atoms forming a group with a lone pair on the nitrogen atom. On exposure to oxygen or water vapour, the nitride surface is hydrolysed to form an amine and a silanol group. Nitride, due to the presence of the amine group, has a higher pH sensitivity of 59 mV for unit change in pH as compared to 30 mV for unit change in pH in oxide. If some of the amines get converted to silanol groups, the pH sensitivity reduces. The sensitivity can be recovered by etching the sample in dilute HF [17] or by a high temperature anneal in nitrogen ambient [14].

3.4 Biosensor

We now proceed to use the pH sensitive EISCAP as a specific biosensor, using the enzymatic hydrolysis of bioanalytes. We will first discuss our studies on triglycerides, often using the short chained tributyrin as a prototype, with the enzyme lipase. We will follow up with the estimation of urea through its hydrolysis in the presence of the enzyme urease. We start with developing the sensor with sample volumes of a few ml and then proceed to miniaturise the sensor with sample volumes in μl. Some applications are discussed in food as well as clinical diagnosis, cross checking the results with existing standard methods.

3.4.1 Triglyceride Sensor

Triglycerides (TGs) are present in the form of fats in food as well as in the human body. TGs in blood are derived from consumed fats in foods or made in the body from other energy sources like carbohyrates. TGs, high-density lipoprotein (HDL) and low-density lipoprotein (LDL) are the constituents of the total cholesterol in human body. Figure 3.11 shows the TG, HDL and LDL and their association with liver, intestine and the blood vessels. Although TGs in human body are acceptable to a certain extent, to deliver energy, high blood TG level is an important risk factor for coronary vascular diseases. Higher TG levels also indicate insulin resistance and very high TGs can cause inflammation of the pancreas (www.webmd.com). The acceptable clinical range for TG level lies between 50 and 150 mg/dL. Table 3.1 shows the TG levels and their severity levels ["What are triglycerides levels and why do they matter?" [Online]. Available: http://triglycerideslevels.org].

Calories that are not consumed by the body are stored as TGs in fat cells and can be converted to energy to meet the demand, but above-normal triglyceride levels or "hyperlipidemia" is indicative of increased risk for heart disease [18]. Presently, total TG levels are estimated spectrophotometrically by estimation of glycerol after elaborate processing of the sample. Even though automation of these methods has

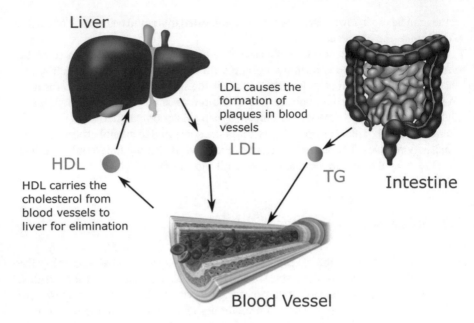

Fig. 3.11 Lipids association with liver intestine and blood vessel (courtesy V.T. Fidal)

Table 3.1 TG levels in blood

Severity level	TG in mg/dL
Normal	<150
Borderline high	150–199
High	200–499
Very high	>500

made them less tedious, the instrumentation is complex and expensive making it essential to develop new and improved biosensors. In the presence of the enzyme lipase, the TG is hydrolysed and there is a change in the pH of the electrolyte in the EISCAP, that shows up as a shift in the C-V plot. Thus, from the shift in the C-V plot, the amount of the TG being hydrolysed can be determined. Determination of TGs from the pH changes caused by the acid produced during enzymatic hydrolysis measured using a field effect based pH sensor have been reported [13, 19–23].

For the use as a biosensor, we choose tributyrin, a short-chained triglyceride (TG), as the bioanalyte and lipase, an enzyme that hydrolyses triglycerides, as the bioreceptor specific to the bioanalyte. TGs are esters composed of one glycerol molecule and three fatty acid molecules and are enzymatically hydrolysed into fatty acids and glycerol. The hydrolysis reaction is given as,

$$\text{Tributyrin} + \text{Water} \overset{\text{Lipase}}{\Leftrightarrow} \text{Glycerol} + \text{Butyric acid}$$

with one mole of TG producing one mole of glycerol and three moles of fatty acid. The butyric acid changes the pH of the electrolyte resulting in a shift in the CV characteristics. The enzyme lipase acts as a catalyst in the hydrolysis of TG referred to as the substrate in biochemistry. The enzymatic hydrolysis is a two-step process: the first step is formation of the enzyme (E) lipase and the substrate (S) TG complex. The formation of this complex is what makes the enzymatic reactions highly specific. In the second step, the complex breaks down to yield the reaction products, leaving the enzyme as it is, as shown below.

$$\text{Enzyme (E)} + \text{Substrate (S)} \rightleftharpoons \text{ES} \xrightarrow{\text{hydrolysis}} \text{Enzyme} + \text{products (P)}$$

The quantities of the substrate, the enzyme and the reaction time—all three play an important role in the enzymatic hydrolysis as described by the Michaelis Menten equation. For a given enzyme concentration, the reaction rate increases linearly with the substrate concentration initially and then saturates. The quantity of enzyme and the time required for the measurement have to be optimised. Enzymes are expensive, so it makes sense to use them in small quantities. Enzymatic reactions are also very specific to temperature, pH, concentration etc. So, a buffer is necessary to provide the right kind of environment for optimal enzyme activity.

The rate of an enzymatic reaction varies linearly with the substrate concentration until the catalytic sites of the enzyme are saturated with the substrate. At saturation, the rate of the reaction becomes independent of the substrate concentrations. Therefore, it is important to optimise various parameters of an enzyme catalysed reaction in the linear region where the specificity of the enzyme to the substrate is maximum. The parameters that were optimised include enzyme concentration, substrate concentration, time of the reaction, the concentration and the pH of the buffer [24] and are given as:

Phosphate buffer: 0.01 M, pH + 7.0.
Enzyme Lipase: 12 mg for 30 ml of Tributyrin solution.
Substrate (Tributyrin) concentration: 25 mM.
Reaction time 30 min.

To calibrate the sensor, we mixed known quantities of the substrate Tributyrin and the enzyme lipase, waited for the appropriate duration for the reaction to take place, and poured it into the Teflon holder. The CV characteristics were measured, and the U_{bias} determined, for different concentrations of the tributyrin. As earlier, the electrolyte used was prepared in a buffer (0.01 M KH_2PO_4/ Na_2HPO_4) solution with KCl (1 M) as the ionic strength adjuster. At the same time, a pH electrode was also used to read the pH value directly for each sample. We thus generate two plots: pH vs the tributyrin concentration and U_{bias} vs pH. Combining the two plots gives us the U_{bias} vs the concentration calibration plot for the sensor as shown in Fig. 3.12.

We can see from the calibration plot that it is linear with a range of 5–20 mM and flattens out at both ends. At higher concentrations of tributyrin (>20 mM), almost all the active sites of the enzyme are already saturated with tributyrin and any further

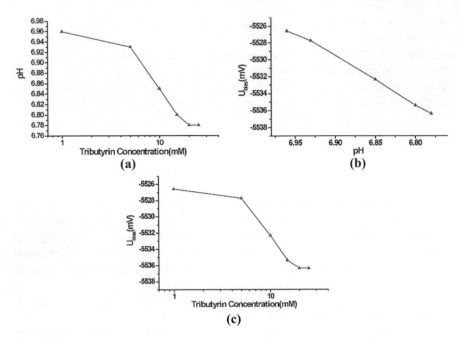

Fig. 3.12 Calibration plot of U_{bias} vs TG concentration for EISCAP **c** generated from **a** pH vs Tributyrin concentration and **b** the U_{bias} measured from the CV plots for the same solutions [a, b courtesy I. Basu, **c** [13]]

increase in the concentration of the bioanalyte does not result in product formation. At low tributyrin concentrations (<0.15 mM), the amount of butyric acid produced is small, as is the resulting change in the pH of the electrolyte. The phosphate buffer in which the electrolyte solution is prepared, inhibits small changes in the pH and hence it cannot be measured. Thus, the lower range limit has more to do with the measurement methodology rather than a device limitation. A lower buffer concentration extends the linear range at lower concentrations, but the CV measurement is found to be unstable and noisy.

The linear range of 0.15 mM to 20 mM (corresponding to 150 mg/dL to 20,000 mg/dL) is adequate for the estimation of triglycerides in blood and falls within the clinical range of Triglyceride concentration. The equivalent of 100 mg/dL is 0.1 mM. Normal fasting levels are generally less than 200 mg/dl. Medications are rarely indicated for levels under 400 mg/dL. High triglycerides (usually over 1000 mg/dl) can cause health problems like pancreatitis, a serious inflammatory condition of the pancreas. Also, human blood does contain a buffer and therefore this is close to a realistic situation.

For a tributyrin sample of an unknown concentration, the previously optimised quantity of lipase is added to the sample and the CV characteristics measured after the optimised time duration. The U_{bias} determined from the CV plot can be used to extract the concentration of tributyrin from the calibration plot 12c). This was verified for several samples and showed good agreement [25].

3.4.2 Triglyceride Estimation in Blood Serum

Due to the presence of buffering action in human blood [26], the analysis of blood as the electrolyte in an EISCAP becomes more complicated than simple triglyceride estimation. Preetha et al. [27] optimised the measurement methodology for quantification of triglycerides in blood serum.

To begin with, Phosphate buffer of different strengths (0.25–10 mM), 5 mM triglyceride (tributyrin) and 1 mg lipase were screened for pH change using a pH meter to optimize the buffer strength for the estimation of triglycerides in blood. Distilled water containing the same concentration of tributyrin and lipase was used as the control. Phosphate buffer at the optimum buffer strength with varying initial pH (7, 6.5 and 6) was mixed with the blood serum and 1 mg of lipase. The pH change due to the hydrolysis of blood triglycerides in these samples was measured using a pH meter. The optimum initial pH was the one which produced an observable pH change. The enzyme activity, at the selected pH and buffer strength, was tested using tributyrin.

To check the potentiometric response of the EISCAP sensor to tributyrin, a control solution of 5 mM tributyrin and 1 mg of lipase was used. A blood sample of 6 ml was collected from volunteers in the age group of 25–40 by trained technicians at the campus hospital, after getting clearance from the Institutional Ethics Committee. Of the 6 ml, 2 ml was used to determine triglyceride levels using a normal laboratory protocol and the remaining 4 ml of blood was used to prepare 'serum' from it by a routine procedure [28]. Blood serum sample (80 µl) was made up to 1.6 ml using phosphate buffer (0.25 M, pH 6) with 1 M KCl as ionic strength adjuster and loaded into the EISCAP. Lipase (1 mg) was added to this solution and mixed well to start the hydrolysis of blood triglycerides. The C–V characteristics of the EISCAP was measured at 2 min intervals over 30 min. For this EISCAP, the measured sensitivity was 55 ± 0.5 mV/pH. Phosphate buffer of concentration 0.25, 0.5 and 1 mM showed detectable pH change during the hydrolysis of the triglyceride (Fig. 3.13). Buffer of 10 mM was too high to show any significant pH change after hydrolysis of the triglyceride. In distilled water, the pH change due to hydrolysis was drastic. In order to ensure reproducible and efficient enzymatic reaction, 0.25 mM buffer was used for analysis of blood. It was found that at very low triglyceride concentration (<1 mM) the buffer strength should be reduced to 0.1 mM in order to obtain an observable pH change. The minimum detectable limit of triglyceride using this sensor was 0.1 mM

Fig. 3.13 Change in pH
with time in different
concentrations of phosphate
buffer for 5 mM tributyrin
and 1 mg lipase [27]

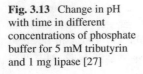

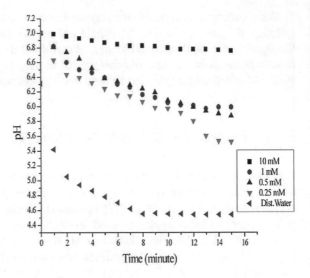

in phosphate buffer (0.1 mM). The pH of the electrolyte changed from 6 to 6.5–
7.5, after adding blood serum and showed a visible change in pH within 10 min on
addition of lipase.

The concentration of triglycerides was measured with the EISCAP and the total
voltage shift ΔU_{bias} over 30 min was found to be 18 mV. A calibration plot of
ΔU_{bias} vs. triglyceride concentration was generated using varying concentrations
of tributyrin (Fig. 3.14). This confirmed that the device responds to the enzymatic
hydrolysis of different tributyrin concentrations in the electrolyte [21]. The amount
of enzyme and the time of reaction were optimized to 1 mg and 15 min respectively,
for the enzyme reaction to be in the linear region.

Fig. 3.14 Tributyrin
standard plot (ΔU_{bias} vs.
tributyrin concentration) [27]

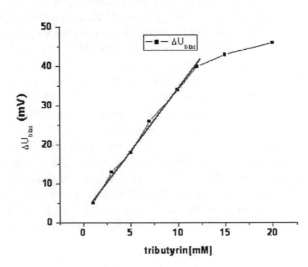

Table 3.2 Triglyceride concentrations in blood serum samples measured using EISCAP

Blood sample	Triglycerides as per the hospital reports (mg/ml)	ΔU_{bias} in mV from C-V curves	Triglycerides levels (mg/ml) from C-V	Error (%)
1	189	23	193	2.12
2	176	19.9	167	5.11
3	259	28.5	239	7.72
4	206	23.2	196	4.85

The optimum response of the EISCAP was obtained when operated at 25 °C in 0.25 mM phosphate buffer (pH 6), 1 M KCl and 1 mg lipase for up to 1 mM concentration of triglycerides.

From the EISCAP measurements, ΔU_{bias} for different blood samples was obtained and from the calibration plot the triglyceride concentrations in blood was determined. The results are tabulated in Table 3.2. In this study no significant difference was observed between the clinical data and the result from the biosensor indicating that the developed EISCAP is a reliable device to analyse blood triglycerides.

3.5 Urea Sensor

Detection and estimation of urea is clinically important, since urea is an important analyte for the diagnosis of diseases such as renal malfunction [29, 30]. The EISCAP can be used for sensing urea through enzymatic hydrolysis in the presence of the enzyme urease [13, 25, 27]. The hydrolysis of urea results in the release of ammonia and carbon dioxide which form aqueous ammonia and carbonic acid in water:

$$\text{Urea} + \text{Water} \xrightarrow{\text{Urease}} \text{Ammonia} + \text{Carbon} - \text{di} - \text{oxide}$$

The formation of twice as much of the strong aqueous ammonia as the weak carbonic acid is reflected in the resulting increase in pH. Hence, pH of the electrolyte solution after this enzymatic reaction shifts towards the basic range.

Urease from Pigeon pea was purified as described by Das et al. [31]. All the optimization studies for the urea sensor were also carried out at 37 °C. Phosphate buffer (KH_2PO_4/Na_2HPO_4) was used for this study. After varying the concentration and pH of buffer (6.75–7.40), it was found that the buffer solution of 0.01 M at pH 7.40 was most effective for carrying out the enzymatic reaction. Hence all the solutions were prepared in 0.01 M phosphate buffer solution (pH 7.40) with 0.5 M KCl added as ionic strength adjuster. To optimize the amount of enzyme (urease), urea solution of 150 mM was prepared in 30 mL buffer (0.5 M KCl in 0.01 M (KH_2PO_4/Na_2HPO_4) and subjected to enzymatic reaction with varying quantities of the enzyme urease. The reactions were monitored with respect to time using a pH meter. The maximum

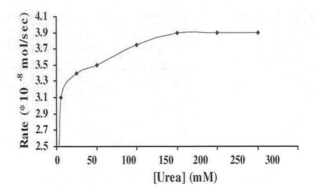

Fig. 3.15 Michaelis–Menten curve [13]

change in pH was obtained with 0.16 mg of urease in 30 mL of 150 mM urea solution. To optimise the urea concentration, enzymatic reactions were then carried out with 0.16 mg urease in 30 mL buffer solution with varying urea concentrations (0 to 250 mM). The reactions were monitored for 30 min using a digital pH meter. After 30 min, the rate of the enzymatic reaction was obtained from the pH value and then plotted against the corresponding urea concentration (Fig. 3.15), which is the Michaelis–Menten curve. From the Michaelis–Menten curve the optimized value of urea concentration obtained is 150 mM. The optimum temperature was found to be 37 °C after carrying out the reaction at 25, 37 and 50 °C. The maximum pH change was seen at 37 °C.

As done earlier for Tributyrin, a calibration plot was generated for urea by combining the two curves, namely: (1) substrate concentration versus pH which is obtained by carrying out the enzymatic reaction on varying concentrations of the substrate (urea), monitoring the change in pH using a digital pH meter and noting down the pH value after the optimised interval of time (i.e. 30 min); (2) U_{bias} versus pH which is obtained by carrying out the C–V measurements using known concentrations of the substrates, after subjecting them to enzymatic reaction for the optimised interval of time and is shown in Fig. 3.16.

With the calibration curves obtained (Fig. 3.16), the device was tested with different unknown concentrations of urea. Various concentrations of the substrate urea, lying in the linear range of the calibration curves were prepared and the optimized amount of the enzyme urease was added. After carrying out the enzymatic reaction for the optimized interval of time, C–V measurements were taken and the corresponding values of U_{bias} was determined. From the calibration curve, the urea concentration corresponding to the measured U_{bias} value was compared with the actual value and found to be in good agreement [13]. Lower detection value for Urea improved considerably from 15 mM seen in Fig. 3.16 to 0.50 mM by the reduction in buffer concentration from 100 to 10 mM using the EISCAPs sensors [25].

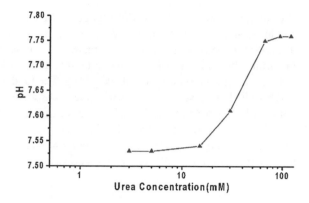

Fig. 3.16 Calibration plot for urea sensor [13]

3.5.1 Measurement of Urea in Blood Serum

For the estimation of urea in blood serum, the EISCAP showed optimum response when operated at 25 °C in 1 mM Tris–HCl buffer (pH 7.4), 0.5 M KCl and 0.5 U urease with a minimum detectable limit of 0.1 mM urea. The response of the EISCAP (post enzyme addition) to known urea concentrations was monitored by doing C–V measurements after complete hydrolysis, which was optimized to 4 min. As expected, it was observed that the shift in C–V increased with higher urea concentrations. As in the case of Triglycerides, a calibration plot was generated with U_{bias} in volts against the urea concentration. The normal level of urea in blood serum is 8–20 mg/dL [32]. Blood urea levels obtained from the EISCAP and the clinical report are given in Table 3.3 and it was found that the biosensor results were as accurate as conventional methods for blood urea.

3.5.2 Measurement of Urea in Milk

Urea $[CO \cdot (NH_2)_2]$, an end product of nitrogen metabolism, has great significance in dairy industry and requires real time monitoring in dairy products during manufacture

Table 3.3 Urea concentrations in blood serum measured using EISCAP

Blood samples	Urea levels as per the clinical test (mg dL^{-1})	U_{Bias} (V) form C-V curves	Concentration of Urea (mg dL^{-1}) from C-V using EISCAP	Error (%)
1	24	−2.920	24.56	2.33
2	27	−2.894	33.81	2.52
3	20	−2.947	14.97	25.15
4	20	−2.936	18.86	5.70

and quality control [33]. Rancidity is caused by a chemical development, which continues until the milk is pasteurized. As butter becomes rancid with time, it breaks down into glycerol and fatty acids—mainly butyric acid. These fatty acids play an important role in its flavor, aroma and texture and measuring the total acid content in butter is an important index of its quality. The total acid content is quantified by a term called the acid value. An acid value greater than 3% and a pH value less than 3.9 qualify butter as rancid [34]. Estimation of the total acid content in butter is a significant application of these pH sensitive EISCAP sensors [25]. Concentration of urea in milk, determined from spectrophotometric measurements, was compared to that obtained from the EISCAP sensor and gave comparable values [27].

3.6 Miniaturised EISCAP

Miniaturisation of the sensors, making them compact and portable, has many advantages. The analysis time can be cut down and the reduced sample volumes require less quantity of other expensive reagents used. Schoning et al. fabricated miniaturized EIS sensors at wafer level by using Si and polymer technologies [35]. They used SU-8 structural polymer layers on EIS structure and by using surface micromachining technique, micro cell EISCAPs were fabricated. They showed biosensors with microcell set-up are similar to those of 'macroscopic' biosensors.

Using Si bulk micromachining technology, it is possible to scale down the sensor dimensions to μm range [36–38]. Instead of the Teflon cell used earlier, 100 μm deep wells are etched into the Si using KOH. These wells act as microreactors and hold the electrolyte, reducing the volume from a few ml to less than a μl. Batch processing of several wafers simultaneously is possible, which can bring down the cost. In addition, multiple EISCAPs on a single silicon die can be fabricated to estimate many bioanalytes simultaneously.

The mini-EISCAPs were fabricated on 1–10 Ω-cm (100) p-type Si wafers with a thickness of 520 μm. Samples were cleaned using standard cleaning techniques to remove dust particles, organic and metallic contaminants. The thickness of SiO_2, used as the mask layer during the KOH etching, has to be high enough to withstand the long KOH wet etching process. Wet oxidation was carried out for 3 h in a dry–wet-dry mode, followed by annealing in N_2 for 20 min, giving a 1 μm thick layer, measured using the ellipsometer. The mask was designed to make 100 devices in a 10×10 array. Buffered Hydrofluoric acid (BHF) with an etch rate of 0.1 μm/min for SiO_2 was used to pattern the windows in oxide. 80% KOH solution (80 gm of KOH in 100 mL of DI water) was prepared, and etching was carried out at 70 °C for 3 h [39]. With a silicon etch rate of 33–35 μm/hour, a depth of 100 μm was expected after wet etching for 3 h. Surface profiler measurements of the pit after the KOH etching indicated a depth of 104 μm. Remaining oxide was removed using BHF or diluted HF solution. Standard RCA cleaning of the sample was done to remove the KOH contaminants. It is very important to perform this RCA cleaning step, to remove all traces of K which can contaminate the processing equipment

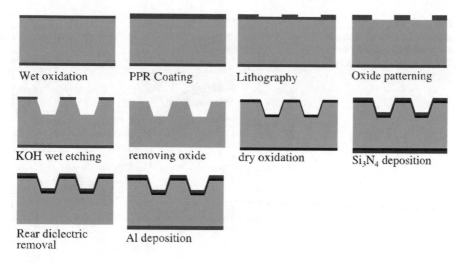

Wet oxidation   PPR Coating   Lithography   Oxide patterning

KOH wet etching   removing oxide   dry oxidation   Si₃N₄ deposition

Rear dielectric   Al deposition
removal

Fig. 3.17 Fabrication steps to make mini-EISCAPs [38]

in subsequent steps. A buffer layer of SiO_2 was grown on the silicon sample with the micro wells. Dry oxidation was carried out for 1 h at 1000 °C followed by N_2 annealing for 20 min. Thickness of the oxide was measured to be 80 nm using an ellipsometer. Silicon nitride layer was deposited at the rate of 5 nm/min on the SiO_2 using Plasma Enhanced Chemical Vapor Deposition (PECVD) followed by annealing in the presence of N_2 at 800 °C for 20 min. Protecting the top surface with PPR, the back oxide and nitride layers were removed using diluted HF. Aluminium was deposited using thermal metal coating system for the back electrical contact. PPR coated on the front side was removed by acetone and cleaned with DI water. Post metallization annealing was carried out at 400 °C for good ohmic contact. The sequence of fabrication steps is illustrated in Fig. 3.17.

The probe station was used as a test bench to take contacts from the mini-EISCAPs. A platinum wire of 150 μm diameter was fitted to the probe head to take contact from the electrolyte. C-V characteristics were measured using the HP4274A LCR meter. To test the pH sensitivity of the device, different pH solutions were made in 10 mM Phosphate buffer in 1 M KCl, as optimised earlier 0.1 to 0.25 μL of the electrolyte, was introduced into the sensor pit using a micro syringe. The sensitivity was measured to be 56 mV/pH.

The first advantage of miniaturisation was seen in the reduced series resistance (520 to 150 Ω) of the mini-EISCAP device compared to the conventional large EISCAP with the teflon cell with a variation in series resistance from 2 kΩ to 300 Ω. This is because of the lower height of the electrolyte column used in the mini-EISCAPs contributing less to the series resistance. The variation in the measured series resistance values is also smaller for mini-EISCAPs, indicating better reproducibility. The lower series resistance allows the reduction of the KCl concentration

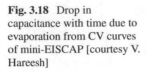
Fig. 3.18 Drop in capacitance with time due to evaporation from CV curves of mini-EISCAP [courtesy V. Hareesh]

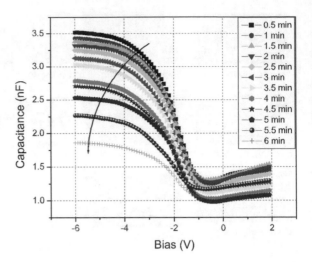

further. However, lower electrolyte volumes—of the order of a fraction of a μL—also led to the evaporation of the electrolyte within a short time. Since the whole surface wall of the well forms the sensor area for the EISCAP, as the electrolyte evaporates, the reduced area of contact for the electrolyte with the sensor surface leads to lower values of C_{max} and therefore the CV plots vary with time as shown in Fig. 3.18. When the micro-well is refilled with fresh electrolyte, the contact area can differ resulting in a different C_{max}. Because of the random variation in contact area, the capacitance changes each time. We could over fill the well but then the additional area outside the well, which can vary, will also contribute to the capacitance. To overcome the variation in the contact area in the mini-EISCAPs, a thicker field oxide layer is grown between the devices using Local Oxidation of Silicon (LOCOS). The additional steps, after forming the silicon reactor by KOH etching, are given below.

A field oxide of 1 μm thickness was grown by wet oxidation at 1000 °C for 4 h followed by annealing in N_2. The field oxide in the active sensor well was etched in BHF solution using the same mask used in forming the wells. The cross section of the device in Fig. 3.19 shows the advantage of the field oxide layer in eliminating the problem of varying contact area. The thickness of the field oxide (t_{fox}) is an order of magnitude higher than that of the gate oxide ($t_{gox} = 80$ nm) and therefore does not contribute to the capacitance. So, as long as the well is overfilled, the capacitance is due to the full active area in the well and remains constant. The pH sensitivity of the device, determined from CV measurements, was found to be 55 mV/pH. The enzyme lipase was immobilised on the surface of the sensor using the protocol described earlier [40].

A 5 mM triglyceride solution was prepared in phosphate buffer and 0.2 μL was put into the lipase immobilized mini-EISCAP device, using a microsyringe. The C-V curves were monitored with time and shows the shift in the C-V characteristics towards the left for Fig. 3.20. As the pH of the electrolyte reduces, due to the formation of butyric acid, it causes this shift in the C-V curves in the depletion region. The

Fig. 3.19 Cross section of the mini-EISCAP with field oxide step after Fabrication [courtesy V. Hareesh]

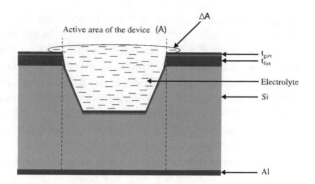

Fig. 3.20 Time evolution of CV curves of Lipase Immobilized mini-EISCAP with Triglyceride solution [38]

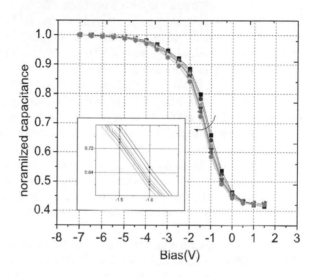

measurements were done every 30 s over 3 min. The shift in the U_{bias} is found to be more initially and as time progresses ΔU_{bias} value reduces. When the hydrolysis reaction is completely over, there is no more shift in the C-V curves. The flatband voltage shift (ΔU_{bias}), as well as the reaction rate for mini-EISCAPs, is much higher than the conventional EISCAP sensors. This can be attributed to the enhanced surface to volume ratio of the miniaturized reactors as compared to that of the conventional EISCAP.

3.6.1 Optimisation of Mini-EISCAPs

The resolution and, in particular, the stability of the mini-EISCAP device can be improved further by process optimisation. A thinner dielectric (insulator) layer in the EISCAP would give a higher capacitance value and a large C-V swing improving the measurement sensitivity and the sensor response. This is especially important when, in field use, the full CV characteristic is not measured using a bridge, but a readout circuit directly measures the shift in CV. A good quality oxide, with a defect free oxide-silicon interface, is important to avoid drift and hysteresis in the CV measurements on the EISCAP devices. It is difficult to control the oxide thickness for thin oxides using conventional furnaces and therefore we switch to Rapid Thermal Oxidation (RTO) to grow a thinner gate oxide [41–43]. This also greatly reduces the thermal budget [44] and the total oxidation duration from 6 h to 11 min [37].

The earlier process steps remain the same, viz a 0.8 μm thick thermal masking oxide grown on a silicon wafer in a wet oxidation quartz furnace (Tempress Systems, Netherlands) following dry–wet- dry sequence for a period of 4 h at 1000 °C. An array of 1500×1500 μm windows were opened in the oxide layer with 2 cm spacing between the windows using buffered hydrofluoric acid (BHF). The exposed silicon was bulk micro machined using semiconductor grade KOH (44.44 wt.%) to a depth of 100 μm for a period of 2 h at 80°C in a constant temperature bath. The wafers were cleaned again using standard Piranha-RCA cleaning procedure to eliminate the K^+ ion contamination and the remaining oxide mask was completely removed from the wafer using BHF solution. A field oxide layer of 0.8 μm thickness was grown thermally at 1000 °C for a period of 4 h. The field oxide plays the dual role of electrically isolating one sensor from the other thereby confining the gate insulator local to the microreactors, as well as reducing the silicon surface roughness that resulted from the KOH etching. The oxide in the microreactors are etched using BHF after patterning it with the same mask used for the KOH etching. After removing the field oxide from the microreactors, a 10 nm gate oxide was grown in a Rapid Thermal Oxidation (RTO) system (Anneal Sys, France) at 1000 °C with an oxygen flow of 50 sccm for 1 min followed by annealing in nitrogen flow of 50 sccm for 3 min. The RTO system was ramped down to 100 °C and the wafers were immediately loaded in a Plasma Enhanced Chemical Vapor Deposition (PECVD) system (Oxford Plasma Technology, UK) to grow a nitride layer of 30 nm thickness at 300 °C for a duration of 8 min with a flow of 25 sccm of Silane, 40 sccm of Ammonia and 100 sccm of Nitrogen. The PECVD system was ramped down to 100°C and the wafer was immediately loaded for annealing at 800 °C in nitrogen ambient for 10 min. The stack of nitride and oxide on silicon is the gate insulating layer of the EISCAP device under study. The RTO grown oxide ensures a good Si- SiO_2 interface and the PECVD deposited nitride is the active pH sensitive layer which is resistant to ion penetration from the electrolyte into the silicon dioxide. the roughness of the RTO grown oxide-Si interface was about 0.2 nm with a good oxide thickness uniformity over the entire wafer [45].

The rear side of the silicon wafers were scratched using a diamond tip for better ohmic contact after removing any residual oxide and nitride and aluminium was thermally evaporated on the back of the wafer for ohmic contact. The chips are diced out using an Ultra slice dicing machine (Ultra Tech Manufacturing, USA) to the designed size of 2×2 cm. The micro machined reactor in each silicon die acts as a miniaturized EISCAP (mini- EISCAP) sensor. The C-V measurements of the mini- EISCAP, after dicing them out from the batch processed wafers were done

Fig. 3.21 Improved C-V swing and slope in the present RTO sample compared to earlier work with conventional dry oxidation [37]

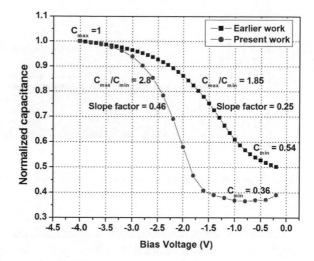

using an Agilent E4980A LCR meter. The sensor chips were loaded on a metal mount, acting as the substrate electrode, and characterized using a Pt wire dipped in the electrolyte as the top electrode as earlier. Phosphate buffer of 10 mM buffer strength and potassium chloride (KCl) of 1 M concentration were used in preparing the electrolyte to characterize the device. Figure 3.21 compares the C-V curves of the sensor with that of our earlier work [27, 46] showing a larger C_{max} to C_{min} ratio with 1.5 times improvement. The dielectric thickness reduction by using the RTO instead of a conventional furnace increases the sensitivity and plays a significant role in improving the performance of the sensor.

We compare the earlier device using conventional thermal dry oxidation with the latter using RTO in Table 3.4. The thickness of the oxide layer, measured using the ellipsometer (Gaertner Scientific Corporation, USA), reduced from 100 to 10 nm. The nitride thickness was also reduced from 70 to 30 nm by reducing the nitride deposition duration in the PECVD system from 20 to 8 min.

Table 3.4 Comparison of process parameters for the EISCSP and the mini-EISCAP (courtesy MS Veeramani)

Parameter	EISCAP	Mini EISCAP
Oxidation furnace	Conventional	RTO
Oxidation time and O_2 flow	1 h, 100 l/hr	60 s, 50 sccm
Annealing time, N_2 flow	20 min, 50 l/hr	3 min, 50 sccm
Oxide thickness	100 nm	10 nm
Nitride deposition time	20 min	8 min
Nitride thickness	70 nm	30 nm
C_{max}/C_{min}	1.85	2.8
Slope factor	0.25	0.46

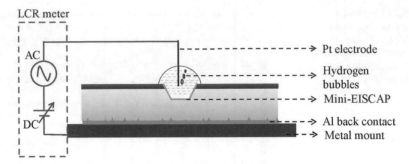

Fig. 3.22 Measurement setup using Pt wire as the top electrode indicating the evolution of bubbles [37]

Choice and integration of the counter electrode is important as the electrode- electrolyte interface requires a stable layer for smooth exchange of charges between them. During the C-V measurements of the mini-EISCAPs, bubbles were seen emanating from the Pt wire counter electrode which eventually developed a black coating [47], resulting in reduced sensitivity. Unless it was cleaned periodically with BHF and aqua regia, the Pt electrode caused time varying random shifts in the C-V characteristics of the sensor [48]. The C-V characteristics of these devices also showed considerable drift with time, much more severe when compared to our earlier work. The C-V measurement setup using Pt wire as the top electrode indicating the evolution of bubbles is shown in Fig. 3.22.

At the electrolyte–electrode interface, the electrons in the electrode and ions in the electrolyte are arranged in a spatial arrangement called the Helmholtz electrical double layer (Fig. 3.23). The charge arrangement develops a potential, referred to as half-cell potential. For a metal electrode, without sufficient ions available on the surface, the half-cell potential of the electrode changes significantly from its equilibrium value when the current density increases beyond the exchange current

Fig. 3.23 Electrical double layer at the Electrode Electrolyte interface [courtesy MS Veeramani]

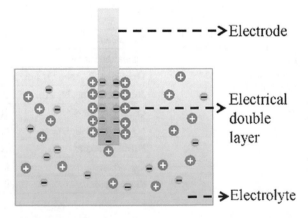

density limit of that electrode, resulting in electrode polarization [48] and exchange of charge is not possible in a polarised electrode [49]. The reduced nitride and oxide thickness in the present samples result in reduced capacitive impedance, increasing the current density beyond the exchange current density limit of the electrode. The polarisation of the Pt electrode causes the C-V drifts over time. Replacing the Pt with gold wire, made the situation worse indicating that the concentration of Pt + and Au + ions was not saturated at the surface of the electrodes affecting the half-cell potential. This problem can be mitigated by using Ag–AgCl electrodes, commonly used as the reference electrode in electrochemistry. When a Ag wire is chloridised, its surface is saturated with AgCl and the equilibrium potential is not affected when current flows. The half-cell potential remains stable, making it a reliable electrode.

The C-V data measured with Pt, Au and chloridised Ag wire electrodes, using a buffer of pH 4.2 as the test electrolyte, is shown in Fig. 3.24. The C-V drift over 10 min was 0.66 mV for the chloridised Ag electrode, 301.9 mV for the Pt electrode and 647.1 mV for the Au electrode. Due to the non-polarisable nature of the chloridised Ag electrode, the drift in C-V with time is much lower when compared to the Pt and Au electrodes. The C-V characteristics of a mini-EISCAP device using chloridised Ag wire as the top electrode is shown in Fig. 3.25 and yielded a sensitivity of 57.3 mV per unit pH change. The chloridised Ag electrode behaves like a conventional Ag–AgCl glass electrode. However, over time, the AgCl coating on the silver wire has a tendency to dissolve in concentrated KCl it is exposed to during the measurements. This may not be an issue for sensors that are of use-and-throw type (limited time usage) and therefore the mini-EISCAPs can be integrated with the chloridised Ag as a solid-state Ag–AgCl electrode [50, 51].

The Pt wire top electrode, used earlier for the mini-EISCAP sensor, is replaced by integrating a reliable thin film Ag–AgCl planar electrode by appropriately treating the silver, deposited on borofloatR glass wafers. First, through holes were etched in the glass wafers by bulk micromachining using a chrome-gold (Cr-Au) layer as the mask.

Fig. 3.24 C-V drift over time with Pt, Au and chloridized Ag electrodes [37]

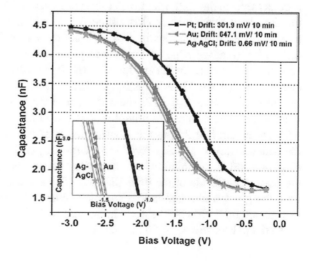

Fig. 3.25 C-V shifts for
solutions with varying pH
using a chloridised Ag wire
electrode [37]

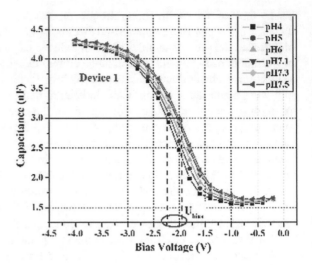

Since the mask has to withstand the etching in concentrated HF for about 60 min, the
Au layer thickness is critical. Hence, after the deposition by e-beam evaporation of a
30 nm Cr layer, to ensure good adhesion, followed by a layer of 120 nm Au on both
sides, the thickness was further increased to 1 μm by electroplating, using gold sulfite
(TSG-250) solution at 60 °C. Using photoresist, circular windows of 1 mm diameter
were opened to etch out the Au and Cr layers. The glass wafers were immersed
in concentrated HF for about 60 min to etch through holes of 3 mm diameter. The
residual Cr-Au-PPR layers were removed after etching the through holes and a thick
layer of 99.9% pure silver was deposited on one side of the micromachined glass
wafer, on a stack of Cr-Au for better adhesion, using e-beam evaporation. Since the
adhesion of silver is poor on glass [52], Cr/Au was initially deposited on the glass
wafer at 120 °C. Figure 3.26 shows the process steps for the fabrication of the thin film
Ag electrode on the glass wafer with the through holes. The glass wafer was diced out
into 2 × 2 cm pieces using an Ultra slice dicing machine. Silver chloride was formed
by placing each of the dies in 3 M KCl solution contained in an electrochemical
cell with a standard Ag–AgCl reference glass electrode as the cathode and the Ag
deposited glass wafer as the anode. After the formation of AgCl, the singulated glass
and silicon dies were bonded using polyimide (PI-2555) or Photoresist (PPR) as an
adhesive as shown in Fig. 3.27. The wafers, after the application of PPR or polyimide,
were soft and hard baked for reliable adhesion. The hardened adhesives do not react
with the electrolyte during the measurements and no leak was observed between
the bonded wafers. The Ag–AgCl layer deposited on the glass wafer acts as the top
electrode and the aluminium layer deposited on the back of the silicon wafer acts as
the bottom electrode of the mini-EISCAP. The silicon microreactor, after bonding
with the glass wafer, can hold a sample volume of 10 μl. C-V measurements on
the bonded mini-EISCAP sensors, integrated with the thin film Ag–AgCl electrode,
yielded a pH sensitivity of 56.5 mV/pH.

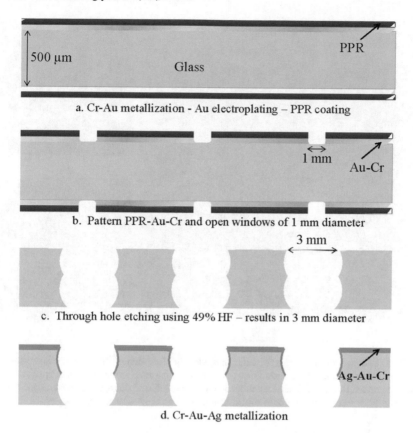

Fig. 3.26 Process steps for the fabrication of thin film Ag electrode on through holes etched glass wafer [courtesy MS Veeramani]

Fig. 3.27 Schematic of mini-EISCAP bonded with thin film Ag–AgCl electrode on glass using PPR/Polyimide [37]

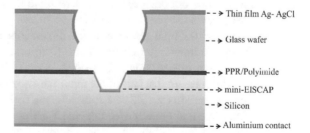

3.7 Mini-EISCAP Triglyceride (TG) Sensor

After bonding the glass and silicon wafers, the enzyme lipase was immobilised by covalent bonding on the nitride surface in the active area of the mini-EISCAP following the protocol developed earlier. The device was immersed in 3 ml of lipase

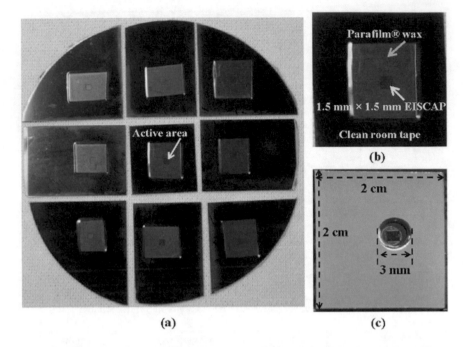

Fig. 3.28 Photograph of **a** a full 3″ wafer immediately after dicing. Picture shows parafilm wax covering the active area of the sensors and the clean room adhesive tape completely covering the wafer, to prevent the slurry, if any, entering into the device during dicing, **b** a diced sensor covered with parafilmR wax and clean room adhesive tape, **c** singulated bonded and enzyme immobilized mini-EISCAP [36]

(1 mg/ml) in Tris– HCl buffer (200 mM, pH 7.2) for 24 h, and finally washed thoroughly with DI water to remove any adsorbed enzyme. Figure 3.28a shows the picture of diced out devices protected with parafilm wax and clean room tape. Figure 3.28b shows the photograph of the singulated sensor and Fig. 3.28c shows the photograph of the final sensor with the dimensions marked. The activity of the enzyme P. cepacia lipase after purification is determined by pNPB assay as reported [40, 53]. The Michaelis constants K_m (shown in Fig. 3.29) for free and immobilised enzyme kinetics measured at half of the maximum enzyme velocity at the maximum substrate concentration ($V_{max}/2$) are found to be 27.97 and 36.48 mM. The specific activity of the free and immobilised enzyme kinetics is 37.8 U/mg and 28.95 U/mg (one enzyme unit 'U' is defined as the amount of enzyme that converts 1 mM of pNPB into product per minute). The immobilised enzyme retained 76.58% of its native activity.

Figure 3.29 shows the rate of the enzymatic reactions of free and immobilised enzyme. The percentage of specific activity of immobilized enzyme relative to that of free enzyme varies from 70 to 80% between different chips that are batch processed.

The TG concentration in blood serum is measured based on the information on the rate of TG hydrolysis used as calibration. Tributyrin, a standard TG, is used in

Fig. 3.29 Free and immobilized enzyme activity plots. Inset shows the percentage of specific activity of immobilized enzyme relative to that of free enzyme [36]

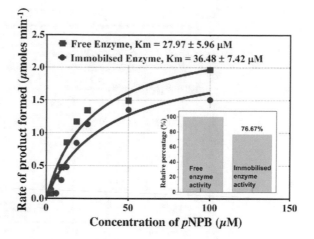

this experiment. Tributyrin concentrations varying from 0.5 to 1.5 mM in steps of 0.25 mM are prepared using Tris–HCl buffer (pH 7.2) and acetonitrile as a co-solvent. The co-solvent plays a major role in mixing the tributyrin homogeneously in the buffer. As discussed earlier, the buffer is necessary to provide a suitable environment for optimal enzyme activity. In our previous work [27] we have optimised the buffer concentration of phosphate buffer to be 0.25 mM at pH 6. However, with this buffer, it was not possible to determine the final pH at a lower TG concentration such as 1 mM of tributyrin that is critical in clinical applications. Hence, in the improved version, we adopted the Tris–HCl buffer. The rate of tributyrin hydrolysis is measured using a pH meter. The inset in Fig. 3.30 shows the measured pH change with time for different tributyrin concentrations. A standard plot (Fig. 3.30) of the average rate of tributyrin concentration vs. [H^+] ion concentration measured between 1 and 2 min of hydrolysis is obtained from the pH vs. time graph measured using a pH meter. The interval of 1 to 2 min is chosen to confine the measurement to the linear region of the enzyme kinetics. The biochemical sensitivity (S_{std}) from the standard plot of the tributyrin is measured to be 0.0481 [H^+] s^{-1} mM^{-1} or 0.00481 pH s^{-1} mM^{-1}.

3.8 Summary

We have discussed the working of the EISCAP sensor and its application to sense triglyceride and urea. Miniaturisation of the sensors has many advantages but also has challenges which can be addressed. Estimated concentration of bioanalytes from the sensors showed good agreement with those from conventional techniques. The readout circuit and the measurement protocol with the mini-EISCAPs will be discussed in the next chapter.

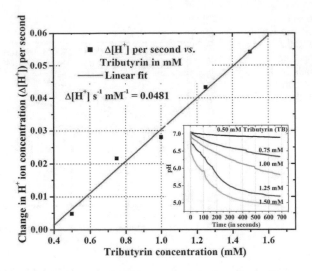

Fig. 3.30 Tributyrin standard plot-average rate of hydrolysis measured at different time instances between 1 and 2 min of the enzymatic reaction is found to be 0.0481 $[H^+]$ s^{-1} mM^{-1}. The inset shows the pH change with time due to hydrolysis of tributyrin of different concentrations with lipase [36]

References

1. Bergveld P (1970) Development of an ion-sensitive solid-state device for neurophysiological measurements. IEEE Trans Biomed Eng BME 17:70–71. https://doi.org/10.1109/TBME.1970. 4502688
2. Siu WM, Cobbold RSC (1979) Basic properties of the electrolyte-SiO2-Si system: physical and theoretical aspects. IEEE Trans Elect Dev 26:1805–1815. https://doi.org/10.1109/T-ED. 1979.19690
3. Poghossian A, Schöning MJ (2020) Capacitive field-effect EIS chemical sensors and biosensors: a status report. Sensors 20:5639. https://doi.org/10.3390/s20195639
4. Schöning MJ, Malkoc Ü, Thust M et al (2000) Novel electrochemical sensors with structured and porous semiconductor/insulator capacitors. Sens Actuat B Chem 65:288–290. https://doi.org/10.1016/S0925-4005(99)00349-4
5. Harame DL, Bousse LJ, Shott JD, Meindl JD (1987) Ion-sensing devices with silicon nitride and borosilicate Grlass insulators. IEEE Trans Elect Dev 34:1700–1707. https://doi.org/10. 1109/T-ED.1987.23140
6. Mikolajick T, Kühnhold R, Schnupp R, Ryssel H (1999) Influence of surface oxidation on the pH-sensing properties of silicon nitride. Sens Actuat B Chem 58:450–455. https://doi.org/10. 1016/S0925-4005(99)00125-2
7. Niu MN, Ding XF, Tong QY (1996) Effect of two types of surface sites on the characteristics of Si3N4-gate pH-ISFETs. Sens Actuat B Chem 37:13–17. https://doi.org/10.1016/S0925-400 5(97)80067-6
8. Vlasov YG, Tarantov YA, Bobrov PV (2003) Analytical characteristics and sensitivity mechanisms of electrolyte-insulator-semiconductor system-based chemical sensors-a critical review. Anal Bioanal Chem 376:788–796
9. Yates DE, Levine S, Healy TW (1974) Site-binding model of the electrical double layer at the oxide/water interface. J Chem Soc Faraday Trans 1 Phys Chem Condens Phases 70:1807–1818. https://doi.org/10.1039/F19747001807

10. Kühnhold R, Ryssel H (2000) Modeling the pH response of silicon nitride ISFET devices. Sens Actuat B Chem 68:307–312. https://doi.org/10.1016/S0925-4005(00)00449-4
11. Van Hal REG, Eijkel JCT, Bergveld P (1996) A general model to describe the electrostatic potential at electrolyte oxide interfaces. Adv Colloid Interface Sci 69:31–62. https://doi.org/10.1016/S0001-8686(96)00307-7
12. Waleed Shinwari M, Jamal Deen M, Landheer D (2007) Study of the electrolyte-insulator-semiconductor field-effect transistor (EISFET) with applications in biosensor design. Microelectron Reliab 47:2025–2057. https://doi.org/10.1016/j.microrel.2006.10.003
13. Basu I, Subramanian RV, Mathew A et al (2005) Solid state potentiometric sensor for the estimation of tributyrin and urea. Sens Actuat B Chem 107:418–423. https://doi.org/10.1016/j.snb.2004.10.038
14. Schöning MJ, Steffen A, Sauke M et al (1995) Improved pH sensitivity of PECVD silicon nitride as gate material for ion-sensitive field-effect transistors (ISFETs). In: Proc. 3rd European Conference Sensors for the Environment, Grenoble, p 55
15. Basu I (2004) Solid state electrochemical biosensor: MS Thesis. Indian Institute of Technology, Madras, India
16. Nicollian EH, Brews JR (2002) MOS (Metal Oxide Semiconductor) Physics and Technology | Wiley. Wiley
17. Liu BD, Su YK, Chen SC (1989) Ion-sensitive field-effect transistor with silicon nitride gate for pH sensing. Int J Electron 67:59–63. https://doi.org/10.1080/00207218908921055
18. Tanne D, Koren-Morag N, Graff E, Goldbourt U (2001) Blood lipids and first-ever ischemic stroke/transient ischemic attack in the Bezafibrate Infarction Prevention (BIP) registry: high triglycerides constitute an independent risk factor. Circulation 104:2892–2897. https://doi.org/10.1161/hc4901.100384
19. Kumar Reddy RR, Basu I, Bhattacharya E, Chadha A (2003) Estimation of triglycerides by a porous silicon based potentiometric biosensor. Curr Appl Phys 3:155–161. https://doi.org/10.1016/S1567-1739(02)00194-3
20. Nakako M, Hanazato Y, Maeda M, Shiono S (1986) Neutral lipid enzyme electrode based on ion-sensitive field effect transistors. Anal Chim Acta 185:179–185. https://doi.org/10.1016/0003-2670(86)80044-7
21. Reddy RRK, Chadha A, Bhattacharya E (2001) Porous silicon based potentiometric triglyceride biosensor. Biosens Bioelectron 16:313–317. https://doi.org/10.1016/S0956-5663(01)00129-4
22. Setzu S, Salis S, Demontis V et al (2007) Porous silicon-based potentiometric biosensor for triglycerides. Phys status solidi 204:1434–1438. https://doi.org/10.1002/pssa.200674378
23. Vijayalakshmi A, Tarunashree Y, Baruwati B et al (2008) Enzyme field effect transistor (ENFET) for estimation of triglycerides using magnetic nanoparticles. Biosens Bioelectron 23:1708–1714. https://doi.org/10.1016/j.bios.2008.02.003
24. Subramanian RV, Basu I, Mathew A et al (2003) Optimization of urease activity for a potentiometric urea biosensor. In: Proc. of IWPSD-2003, Intl. Workshop on Physics of Semiconductor Devices. Chennai, India, pp 734–736
25. Mathew A, Pandian G, Bhattacharya E, Chadha A (2009) Novel applications of silicon and porous silicon based EISCAP biosensors. Phys status solidi 206:1369–1373. https://doi.org/10.1002/pssa.200881084
26. Salenius P (1957) A study of the ph and buffer capacity of blood, plasma and red blood cells. Scand J Clin Lab Invest 9:160–167. https://doi.org/10.3109/00365515709101216
27. Preetha R, Rani K, Veeramani MSS et al (2011) Potentiometric estimation of blood analytes—triglycerides and urea: comparison with clinical data and estimation of urea in milk using an electrolyte-insulator-semiconductor-capacitor (EISCAP). Sens Actuat B Chem 160:1439–1443. https://doi.org/10.1016/j.snb.2011.10.008
28. Varley H (1958) Practical clinical biochemistry. Interscience Publishers
29. Chen JC, Chou JC, Sun TP, Hsiung SK (2003) Portable urea biosensor based on the extended-gate field effect transistor. Sens Actuat B Chem 91:180–186. https://doi.org/10.1016/S0925-4005(03)00161-8

30. Lakard B, Herlem G, Lakard S et al (2004) Urea potentiometric biosensor based on modified electrodes with urease immobilized on polyethylenimine films. Biosens Bioelectron 19:1641–1647. https://doi.org/10.1016/j.bios.2003.12.035

31. Das N, Prabhakar P, Kayastha AM, Srivastava RC (1997) Enzyme entrapped inside the reversed micelle in the fabrication of a new urea sensor. Biotechnol Bioeng 54:329–332. https://doi.org/10.1002/(SICI)1097-0290(19970520)54:4%3c329::AID-BIT5%3e3.0.CO;2-M

32. Ahuja T, Mir IA, Kumar D, Rajesh, (2008) Potentiometric urea biosensor based on BSA embedded surface modified polypyrrole film. Sens Actuat B Chem 134:140–145. https://doi.org/10.1016/j.snb.2008.04.020

33. Trivedi UB, Lakshminarayana D, Kothari IL et al (2009) Potentiometric biosensor for urea determination in milk. Sens Act B Chem 140:260–266. https://doi.org/10.1016/j.snb.2009.04.022

34. Clb A, Nb G, Mancini Filho J (2001) Características De Identidade, Qualidade E Estabilidade Da Manteiga De Garrafa: Parte I - Características De Identidade E Qualidade. Ciência e Tecnol Aliment 21:314–320. https://doi.org/10.1590/s0101-20612001000300011

35. Schöning MJ, Näther N, Auger V et al (2005) Miniaturised flow-through cell with integrated capacitive EIS sensor fabricated at wafer level using Si and SU-8 technologies. Sens Actuat B Chem 108:986–992. https://doi.org/10.1016/j.snb.2004.12.029

36. Veeramani MS, Shyam KP, Ratchagar NP et al (2014) Miniaturised silicon biosensors for the detection of triglyceride in blood serum. Anal Methods 6:1728–1735. https://doi.org/10.1039/c3ay42274g

37. Veeramani MS, Shyam P, Ratchagar NP et al (2013) A miniaturized pH sensor with an embedded counter electrode and a readout circuit. IEEE Sens J 13:1941–1948. https://doi.org/10.1109/JSEN.2013.2245032

38. Vemulachedu H, Fernandez RE, Bhattacharya E, Chadha A (2009) Miniaturization of EISCAP sensor for triglyceride detection. J Mater Sci Mater Med 20:229–234. https://doi.org/10.1007/s10856-008-3534-y

39. Noor MM, Bais B, Majlis BY (2002) The effects of temperature and KOH concentration on silicon etching rate and membrane surface roughness. In: IEEE International Conference on Semiconductor Electronics, Proceedings, ICSE, pp 524–528

40. Fernandez RE, Bhattacharya E, Chadha A (2008) Covalent immobilization of Pseudomonas cepacia lipase on semiconducting materials. Appl Surf Sci 254:4512–4519. https://doi.org/10.1016/j.apsusc.2008.01.099

41. Kireev VY, Tsimbalov AS (2001) Rapid thermal processing: a new step forward in microelectronics technologies. Russ Microelect 30:225–235. https://doi.org/10.1023/A:1011346427728

42. Nulman J, Krusius JP, Gat A (1985) Rapid thermal processing of thin gate dielectrics. oxidation of silicon. IEEE Elect Device Lett 6:205–207. https://doi.org/10.1109/EDL.1985.26099

43. Tung NC, Caratini Y, Caratini Y (1986) Rapid thermal oxidation of silicon for thin gate dielectric. Electron Lett 22:694–696. https://doi.org/10.1049/el:19860475

44. Hollauer C (2007) Modeling of thermal oxidation and stress effects. PhD thesis, Vienna University of Technology http://www.ub.tuwien.ac.at/englweb/

45. Mur P, Semeria MN, Olivier M et al (2001) Ultra-thin oxides grown on silicon (1 0 0) by rapid thermal oxidation for CMOS and advanced devices. Appl Surf Sci 175–176:726–733. https://doi.org/10.1016/S0169-4332(01)00081-2

46. Mohanasundaram SV, Mercy S, Harikrishna PV et al (2010) Packaged bulk micromachined triglyceride biosensor. In: Reliability, Packaging, Testing, and Characterization of MEMS/MOEMS and Nanodevices IX. SPIE, p 75920G

47. Yazici B (1999) Hydrogen evolution at platinum (Pt) and at platinized platinum (Ptz) cathodes. Turkish J Chem 23:301–308

48. Taing M (2009) Characterisation and fabrication of a Multiarray Biosensor. Griffith University, Queensland, Australia

49. Dankers TEH (1996) Electrische spanningsvariaties in oppervlakte-electroden: Master's Thesis. University of Amsterdam

50. Huang IY, Huang RS, Lo LH (2003) Improvement of integrated Ag/AgCl thin-film electrodes by KCl-gel coating for ISFET applications. Sens Actuat B Chem 94:53–64. https://doi.org/10.1016/S0925-4005(03)00326-5
51. Smith RL, Scott DC (1986) An integrated sensor for electrochemical measurements. IEEE Trans Biomed Eng BME 33:83–90. https://doi.org/10.1109/TBME.1986.325881
52. Benjamin P, Weaver C (1961) The adhesion of evaporated metal films on glass. Proc R Soc London Ser A Math Phys Sci 261:516–531. https://doi.org/10.1098/rspa.1961.0093
53. Shirai K, Jackson RL (1982) Lipoprotein lipase-catalyzed hydrolysis of p-nitrophenyl butyrate. Interfacial activation by phospholipid vesicles. J Biol Chem 257:1253–1258
54. Piskorski K, Przewłocki H (2009) LPT and SLPT measurement methods of flat-band voltage (VFB) in MOS devices. J Telecomm Inf Technol 4:76–82

Chapter 4
Readout Circuit and Measurement Protocol for Mini-EISCAPs

The readout electronics is developed for an Electrolyte Insulator Semiconductor capacitor (EISCAP) sensor. The sensor detects the presence of bioanalytes from changes in the pH of the electrolyte through shifts in the capacitance–voltage characteristics. An EISCAP relaxation oscillator is embedded in a successive approximation analog to digital conversion algorithm to give a digital readout of the pH of the test solution. Further, a compact readout system is developed and implemented in a Programmable System on Chip (PSoC®). A measurement protocol using the miniaturised EISCAP devices is described, incorporating calibration to account for sensor variations. Testing is done using blood serum samples to estimate the triglyceride concentration within the clinical range of 50 to 150 mg/dL and compared with results from standard clinical measurement techniques.

In EISCAPs there is a shift in the flat-band voltage, and hence the capacitance–voltage (CV) characteristics, with a change in the pH of the electrolyte as shown in Fig. 4.1. Enzymatic hydrolysis of triglycerides (TG) produces butyric acid and therefore the EISCAP, immobilized with the enzyme lipase, can be used as a specific biosensor for TG. Using micromachining, and adding an embedded Ag–AgCl counter electrode, the sensor was miniaturised for ease of use and portability. Typically, CV measurements are done in the laboratory using a bridge and is time consuming. Developing a compact readout circuit for measuring the flat-band voltage of the EISCAP [1, 2] furthers the ease of measurements and portability. The measurement protocol is optimized for the mini-EISCAP, with the readout circuit, for pH sensing as well as estimation of triglycerides in blood serum [3].

4.1 Principle of Flat-Band Voltage Measurement

Typical C-V curves for an EISCAP, which is a nonlinear capacitor, with two electrolytes of pH values pH_1 and pH_2 ($pH_2 > pH_1$) are shown in Fig. 4.2.

© The Author(s), under exclusive license to Springer Nature Switzerland AG 2021
E. Bhattacharya, *Biosensing with Silicon*, SpringerBriefs in Materials,
https://doi.org/10.1007/978-3-030-92714-1_4

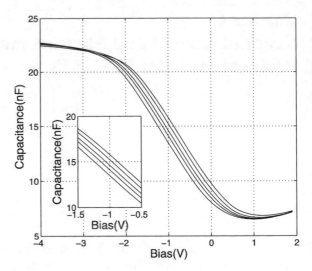

Fig. 4.1 Measured C-V plots for EISCAP with the electrolyte pH varying from 3 to 11 in steps of 2 [2]

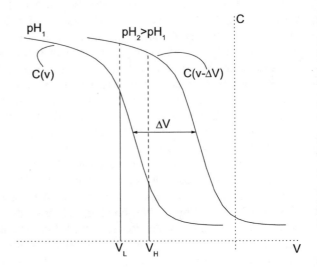

Fig. 4.2 Typical EISCAP C-V curves for electrolytes with different pH [2]

For the EISCAP, when the electrolyte pH value is pH_1 and the capacitance–voltage characteristic $C(v)$ then, the time taken to charge the EISCAP from an initial voltage V_A to a final voltage V_B with a constant current source I, is given by:

$$t_1 = \frac{1}{I} \int_{V_A}^{V_B} C(v) dv \qquad (4.1)$$

We assume, for simplicity, that t_1 is also the time required to discharge the EISCAP from an initial voltage V_B to a final voltage V_A, using the same current source. To measure t_1, the charging and discharging of the EISCAP is incorporated into a

Fig. 4.3 Simplified schematic of the EISCAP relaxation oscillator [2]

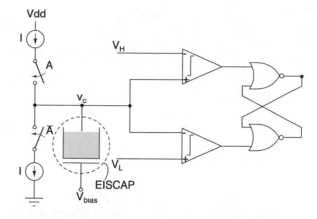

relaxation oscillator as shown in Fig. 4.3. Two comparators detect the voltage at the top plate of the EISCAP. As soon as V_C exceeds V_H, the current source pulling V_C up is disabled and the pull-down current is enabled. The opposite happens when V_C goes lower than V_L, ensuring that V_C always stays between V_H and V_L. The bottom plate of the EISCAP is connected to the variable voltage V_{bias}. The motivation for this will become clear in the discussion to follow. From Eq. (4.1), the frequency f_1 of the relaxation oscillator of Fig. 4.3 is:

$$f_1 = \frac{I}{2}\left[\int_{V_L}^{V_H} C(v_c - V_{bias})dv_c\right]^{-1} \tag{4.2}$$

When the pH of the electrolyte changes to pH_2, the C-V characteristic becomes $C(V - \Delta V)$, where ΔV is given by Eq. (4.3) derived from the ideal Nernst Eq. [4.4]:

$$\Delta V = 2.303\frac{kT}{q}(pH_2 - pH_1) \tag{4.3}$$

From Fig. 4.3, it is seen that, for voltages in the range $V_L < V_c < V_H$, the capacitance becomes larger than what it was when the pH of the electrolyte was pH_1. Thus, the frequency of the relaxation oscillator will reduce to f_2:

$$f_2 = \frac{I}{2}\left[\int_{V_L}^{V_H} C(v_c - \Delta V - V_{bias})dv_c\right]^{-1} \tag{4.4}$$

From Eqs. (4.2) and (4.4), we see that f_2 can now be made equal to f_1, if V_H and V_L are changed to $V_H + \Delta V$ and $V_L + \Delta V$ respectively. Thus, the protocol to determine the flat-band voltage shift (ΔV) in the EISCAP for solutions with two different pH values can be formulated as follows.

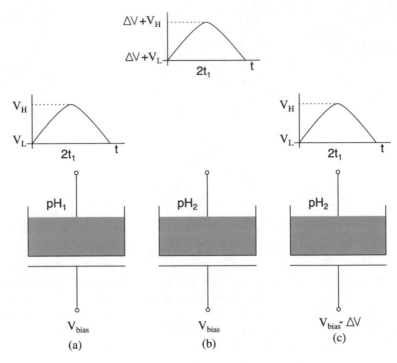

Fig. 4.4 Measurement of flat-band voltage with the relaxation oscillator [2]

For an electrolyte with a known $pH = pH_1$ in the EISCAP, the frequency f_1 of the relaxation oscillator is measured. The voltage waveform at the EISCAP top plate is shown in Fig. 4.4a. Values of V_H and V_L must be chosen such that they lie in the depletion region (the region with the highest slope) of the C-V curve.

Next, the measurement is done with the electrolyte of $pH_2 > pH_1$ - the frequency of the oscillator is now $f_2(< f_1)$. V_H and V_L are increased by the same amount until f_2 equals f_1. The change in V_H (and V_L) is the desired ΔV and the waveform is as shown in Fig. 4.4b. For an easier implementation, one could achieve the same result leaving V_H and V_L unchanged and lowering the potential V_{bias} of the bottom plate of the capacitor by ΔV, as shown in Fig. 4.4c. The value of ΔV is the flat-band voltage of the EISCAP.

4.2   Measurement of pH

If pH_2 is the unknown pH of a test solution it can, in principle, be determined from the known pH_1 and the measured ΔV values using Eq. (4.3). However, this equation is derived assuming a theoretical sensitivity of 59.3 mV/pH for nitride surfaces. In practice, the pH sensitivity can vary anywhere between 40 and 55 mV/pH depending on the quality of the nitride. A more general formula for ΔV is:

$$\Delta V = 2.303 \frac{\alpha k T}{q} \left(pH_2 - pH_1 \right) \tag{4.5}$$

Where α, a dimensionless sensitivity parameter with a value less than one, depends on the quality of the insulator surface. In addition, since the flat-band voltage is process dependent and can vary from run to run [2, 5] and even in different EISCAPs in the same run, it is difficult to ensure a priori that the chosen values of V_H and V_L puts the EISCAP in the depletion region of the C-V characteristic. To address these issues for the digital readout, the relaxation oscillator with the EISCAP is embedded in a successive approximation (SAR) analog-to-digital conversion algorithm as discussed below. A microcontroller with a suitable interface logic is used to accomplish all the tasks, as shown in Fig. 4.5. The potential of the lower plate of the EISCAP is set from the microcontroller using two 8-bit Digital-to-Analog Converters (DAC). DAC_1 and DAC_2 are referred to as the coarse and fine DACs respectively.

From Fig. 4.5, it is seen that,

$$V_{bias} = V_{coarse} \left(1 + \frac{R_1}{R_2} \right) - V_{fine} \frac{R_1}{R_2} \tag{4.6}$$

R_1 and R_2 are chosen so that V_{bias} changes by 2 mV for every LSB change in the input of the fine DAC, and by about 30 mV for every LSB change of the coarse DAC. The steps to measure the pH of the analyte follow [2].

1. With the fine DAC code is set to zero initially, an electrolyte with pH $= 2$ (pH1) is used in the EISCAP. The frequency of the EISCAP oscillator is measured for the minimum and maximum values of the input code of the coarse DAC. The highest

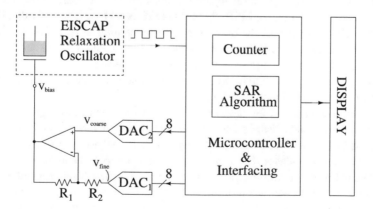

Fig. 4.5 Block diagram of the pH readout instrument [2]

and lowest values of the DAC output should put the EISCAP in accumulation and inversion, resulting in the lowest (f_L) and highest (f_H) oscillation frequency respectively. Then, a successive approximation algorithm is used to determine the coarse DAC code needed to achieve an oscillation frequency of about (f_L + f_H)/2. This ensures that the EISCAP is biased in the middle of the depletion region, independent of the flat-band voltage.

2. The frequency of the EISCAP f_1 is found from the number of oscillator cycles in a 500 ms interval.

3. A second reference electrolyte with pH = 11 is also used, giving a lower frequency of oscillation as in Eq. (4.4), since there are two unknown parameters 'α' and 'pH$_2$' in Eq. (4.5). The fine DAC bits are adjusted via successive approximation to bring back the oscillation frequency to f_1. This is the voltage shift (ΔV) in the C-V characteristics of the EISCAP for a pH change of 9 and therefore provides the calibration.

4. Similarly, for the sample analyte of unknown pH and hence an unknown frequency of oscillation, the fine DAC code required to bring the frequency of oscillation to f_1 is determined using successive approximation.

5. From the fine DAC codes obtained in the two steps above, pH$_2$, the unknown pH of the analyte is determined from Eq. (4.7) and displayed on the LCD screen of the readout.

$$pH_2 = \left(\frac{\Delta V_{test}}{\Delta V_{ref}}\right)\Delta pH_{ref} + pH_1 \qquad (4.7)$$

ΔV_{ref}: ΔV between pH = 2 and pH = 11.
ΔV_{test}: ΔV between pH = 2 and test pH solution.
ΔpH_{ref}: difference between the pH values of the reference solutions. In this case it is 9 since we use pH =2 and 11 as the reference solutions.

The successive approximation algorithm, illustrated for the case of the fine DAC, operates as follows. First, the MSB of the fine DAC is set to '1', and all other bits are set to '0'. The frequency of the oscillator (f_2) is compared to f_1. If $f_2 < f_1$, the next significant bit is set to '1'. If $f_2 > f_1$, the MSB is set to '0' and the next significant bit is set to '1'. This process continues until all the bits are set.

Good agreement was found between the results using the readout circuit and those from the C-V measured using a bridge.

As mentioned earlier, fabrication process variations result in a significant range in the flat-band voltage values and the pH sensitivity of the EISCAP. To test the robustness of our readout circuit in the face of EISCAP variations, nine EISCAP devices were used as sensors to measure the pH of standard test solutions. The results are shown in Fig. 4.6. Including the calibration techniques in the circuit makes the pH measurement robust even when different EISCAPs are used. The variation seen in the measured pH was attributed to the finite resolution of the DAC, and the nonlinear shift in the CV plot with pH for the EISCAP.

Fig. 4.6 Response of the
readout circuit with nine
different EISCAP sensors [2]

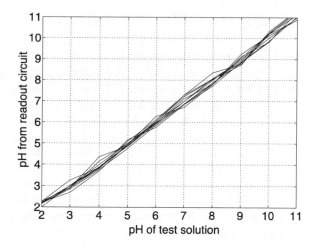

Fig. 4.7 pH values
measured after enzymatic
hydrolysis, for different
initial concentrations of
Tributyrin [2]

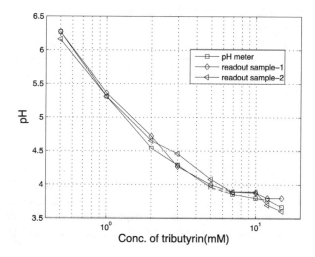

The EISCAP sensor and the readout circuit were used to estimate the concentration of tributyrin through its enzymatic hydrolysis, in the presence of lipase. The butyric acid produced results in a change in the pH of the electrolyte. Figure 4.7 shows the measured pH, after the completion of the hydrolysis reaction, for different concentrations of tributyrin. Results using two EISCAP sensors show a good match to the readings obtained using a commercial pH meter.

4.3   System-On-Chip Implementation

The pH readout system for the mini-EISCAP was implemented using a mixed signal system-on-a-chip (SoC) from Cypress Semiconductors (PSoCR) [6] with minimal external components [1, 3]. The SoC incorporates analog digital mixed signal processing blocks and the firmware. The SoC implementation addresses the sensor nonidealities due to process variations and also provides a digital readout for the pH of an unknown electrolyte under test. Figure 4.8 shows the block circuit schematic used in this work. The mini-EISCAP sensor incorporated in an op-amp based relaxation loop generates the mini-EISCAP clock. The frequency of the mini-EISCAP clock is measured by counting the number of system clock cycles in one EISCAP clock cycle. The output of the frequency measuring logic is latched for every cycle of the mini-EISCAP pulse. The frequency count is fed to the status register that provides easy access through a programming interface. The mini-EISCAP relaxation oscillator that is embedded in a successive approximation analog to digital conversion feedback loop (SAFL) digitizes the measured V and also addresses the validity of Eq. (4.5).

Equation (4.8) describes the bottom plate voltage (V_{bias}) of the mini-EISCAP that was set from the firmware using two 8 bit Digital-to-Analog-Converters (DACs). DAC_1 and DAC_2 are referred to as the fine and coarse DAC respectively. The V_{bias} in Fig. 4.8 is,

$$V_{bias} = V_{coarse}\left(1 + \frac{R_2}{R_1}\right) - V_{fine}\left(\frac{R_2}{R_1}\right) \qquad (4.8)$$

The calibration of the sensor is discussed next. The sensor was biased in accumulation and inversion regions (at the edge of the depletion region) after placing the electrolyte with pH_4 in the sensor. The mini-EISCAP relaxation oscillator produces frequencies of oscillation f_L and f_H when biased in these constant capacitance regions of the C-V characteristics. To achieve this, the fine DAC input code was reset to all

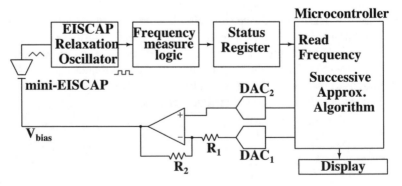

Fig. 4.8 Block diagram of the SoC implementation for the pH measurement system [1]

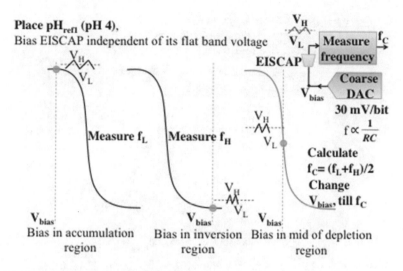

Fig. 4.9 First calibration step shown using C-V curves (courtesy MS Veeramani)

'0's and the coarse DAC input code was set to all '1's followed by all '0's so that the sensor was biased at the extreme ends of the C-V curve. To bias the mini-EISCAP with the help of the coarse DAC in the middle of the depletion region, independent of the flat-band voltage, the centre frequency f_c was calculated as an arithmetic mean of f_L and f_H. The coarse DAC input code was then adjusted using successive approximation algorithm (SAFL), driven by the firmware, till it achieved f_c frequency of oscillation. This was the first step in calibration (shown in Fig. 4.9).

In the second calibration step, shown in Fig. 4.10, the electrolyte pH_8 was placed in the sensor and the frequency of oscillation reduced. Now, the fine DAC input code was adjusted with the help of successive approximation algorithm (SAFL) to bring back the frequency of oscillation to f_c. The fine DAC input code that results in the oscillation frequency f_c corresponds to the voltage shift (V_{ref}) in the EISCAP C-V characteristics for a change in pH of 4. V_{bias} changes by 2 mV for every LSB change in the input code of the fine DAC and by 24 mV for every LSB change in the input code of the coarse DAC and hence, R_1 and R_2 are chosen accordingly.

4.3.1 Measurement of pH

To determine the unknown pH ($=pH_x$) of a test electrolyte, it was placed in the mini-EISCAP sensor. The successive approximation algorithm (SAFL) again adjusts the fine DAC code required to bring back the frequency of oscillation to f_c. The new input code corresponds to the voltage shift (V_{test}) in the mini-EISCAP C-V characteristics. The unknown pH_x was then determined using Eq. (4.9).

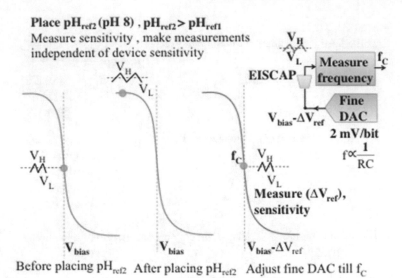

Fig. 4.10 Second calibration step shown using C-V curves (courtesy MS Veeramani)

$$pH_x = pH_8 - \Delta pH_{ref}\left(\frac{\Delta V_{test}}{\Delta V_{ref}}\right) \tag{4.9}$$

$$\Delta V_{ref} = 0.0592\alpha(pH_8 - pH_4)$$

$$\Delta V_{test} = 0.0592\alpha(pH_8 - pH_x)$$

The result of the test pH_x was displayed on the LCD screen. The measurement mode using the mini-EISCAP C-V characteristics, that is used in the pH readout, is shown in Fig. 4.11.

The successive approximation algorithm works as follows. After calibrating the sensor for the centre frequency fc, the MSB of the DAC is set to '1' and the rest all set to '0'. The measured frequency of oscillation (*fmeasure*) is compared to fc. If *fmeasure* < fc, the next significant bit is set to '1', else if *fmeasure* > fc, the MSB is set to '0' and the next significant bit is set to '1'. This is repeated until it covers all the bits of the DAC input code.

The pH readout measurements were performed on the mini-EISCAP sensors after calibrating them using the two reference electrolytes pH 4 and pH 8. The nominal frequency of the EISCAP relaxation oscillator was set as 10 kHz and the PSoC® system clock frequency is set as 10 MHz. The test electrolytes chosen varied from pH 4.5 to pH 7 in steps of 0.5. Table 4.1 compares the pH of the test solution measured with the pH readout system against the Eutech pH meter measurements for the sensors S1, S2 and S3. Figure 4.12 shows the photograph of the designed pH readout system connected with the mini- EISCAP.

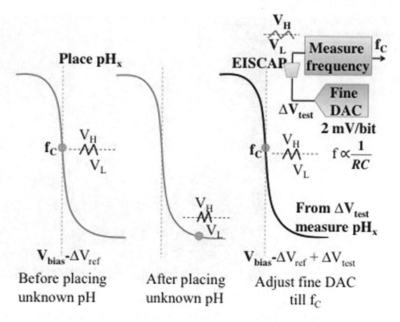

Fig. 4.11 Measurement mode shown using C-V curves (courtesy MS Veeramani)

	S. No	pH meter	Readout System	Error (%)
Table 4.1 Comparison of the readings from the pH meter and the readout circuit [1]	1	4.0	Calibration	–
	2	4.5	4.40	2.2
	3	5.0	5.10	2.0
	4	5.5	5.65	2.7
	5	6.0	6.00	0.0
	6	6.5	6.57	1.0
	7	7.0	7.05	0.7
	S	8.0	Calibration	–

A flow chart of the pH measurement protocol is given in Fig. 4.13.

4.4 Readout Protocol for Estimating TG with EISCAP

The principle used in the readout system to determine the TG in blood serum is similar to the one used for pH sensing. The mini-EISCAP sensor was embedded in an oscillator loop whose frequency (f) gets modulated with changes in the pH of the blood serum due to the enzymatic hydrolysis of the TG. A successive approximation

Fig. 4.12 Photograph of the
pH readout system with the
mini-EISCAP [1]

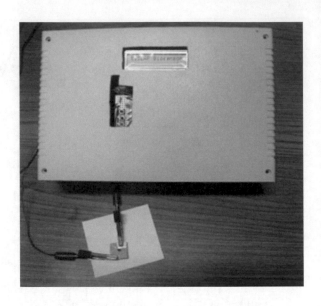

algorithm, along with the oscillator, determines the flat band shift (due to the pH change) of the mini-EISCAP by measuring this frequency (f). The shift in the flat band was correlated to the amount of TG in the blood serum.

In the pH readout system, pH 4 and pH 8 were used as reference electrolytes to calibrate the sensor against process variations. Since the TG measurement was done on an enzyme immobilized sensor, the standard pH reference solution of pH 4 cannot be used as the enzyme could be deactivated at this pH. Hence, the reference pH solutions for the TG readout system were chosen to be pH 6 and pH 8.5 to calibrate the sensor against process variations using Eq. (4.10), where α is the dimensionless parameter with a value less than 1 that depends on the quality of the insulator.

$$\Delta V_{ref} = 59.2\alpha(pH_{8.5} - pH_6) \tag{4.10}$$

The readout circuit and the algorithm for the measurement of TG in blood serum were implemented using a mixed signal embedded chip from Cypress Semiconductors (PSoC3R, CY8C3866AXI-040) [6] ["Cypress semiconductors psoc3 datasheet," Cypress Semiconductors Pvt. Ltd., San Jose, CA, 2009. http://www.cypress.com/go/psoc3] with minimal external components on a compact custom-made printed circuit board of size 11 × 6 cm. Figure 4.14 shows the block circuit schematic of the TG measurement system used. The EISCAP sensor incorporated in an op-amp based relaxation loop generates the EISCAP clock. The frequency of the EISCAP clock is measured by dividing the EISCAP clock by 32 and counting the number of system clock cycles in one EISCAP divide-by-32 clock cycle. The counter that counts the system clock cycles is a 21-bit counter. In this averaging technique, the LSB 5 bits of the counter are skipped and the MSB 16 bits are registered for every cycle of the EISCAP divide-by-32 pulse. In the pH measurement system, the averaging was done

pH measurement procedure

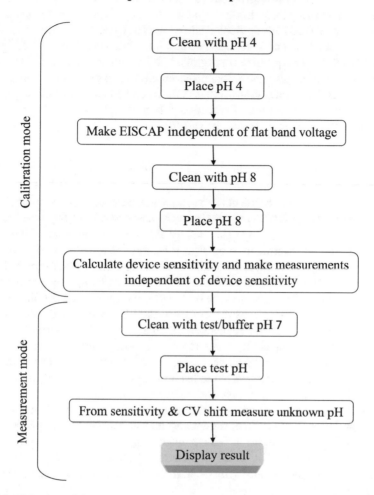

Fig. 4.13 Flow chart of the pH measurement protocol (courtesy M.S. Veeramani)

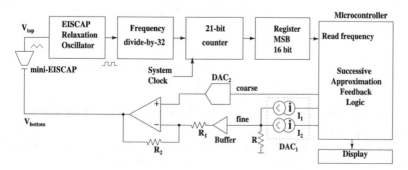

Fig. 4.14 Block diagram of the TG measurement system [1]

by adding the number of system clock cycles in every cycle of EISCAP for 50 cycles. This way of quantizing the number of system clock cycles adds up the quantization noise by 50 times i.e. for every clock cycle of the EISCAP. The modified technique reduces the quantization noise to one time and this noise is further reduced by skipping the LSB 5 bits. The quantized or averaged 16-bit data (frequency count) from the counter is fed to the status register that provides easy access through a programming interface. The EISCAP relaxation oscillator, that is embedded in a successive approximation analog to digital conversion feedback loop (SAFL), digitises the measured ΔV_{ref} (Eq. 4.10) and also addresses its validity.

$$V_{bottom} = V_{coarse}(1 + \frac{R_2}{R_1}) - V_{fine-combined}(\frac{R_2}{R_1}) \qquad (4.11)$$

Equation (4.11) describes the bottom plate voltage (V_{bottom}) of the mini-EISCAP that is set from the (microcontroller) firmware using two 8-bit Digital-to-Analog-Converters (DACs), an op-amp (gain set by R_1 and R_2) and buffer (unity gain). DAC_1 and DAC_2 are referred to as the fine and coarse DAC respectively. DAC_1 is a parallel combination of two fine current DACs I_1 and I_2. The voltage drop $R(I_1 + I_2)$ is the fine DAC_1's output voltage and the buffer prevents $R_{1\,from}$ loading the DAC_1. The V_{bottom} seen in Fig. 4.14, changes by 0.3125 mV for every LSB change in the input code of the fine DAC_1 I_1, by 2.5 mV for every LSB change in the input code of the fine DAC_1 I_2, and by 32 mV for every LSB change in the input code of the coarse DAC_2 and therefore R, R_1 and R_2 have to be chosen accordingly. The calibration and measurement modes are discussed next using the EISCAP C-V characteristics.

The calibration and measurement modes for the estimation of Triglycerides using the EISCAP C–V characteristics are shown in Fig. 4.15.

The sensor is calibrated by biasing it in the middle of the C–V curve (where the EISCAP produces $f = f_{mid}$ clock frequency) and by measuring the pH sensitivity (ΔV_{ref} / 2.5) with pH 6 and pH 8.5 electrolytes using Eq. (4.10).

The biochemical sensitivity from the standard plot, discussed earlier (Sstd after 1 min duration) and the pH sensitivity (S, in mV/ΔpH) of the device, calculated from (10), are given as:

$$S_{std\_1min} = 0.028\,pH/mM \qquad (4.12)$$

$$S = \frac{\Delta V_{ref}}{2.5} \qquad (4.13)$$

The blood serum sample, for which the TG concentration is to be determined, is placed in the sensor. The initial voltage and the voltage after 1 min to bring back the frequency of oscillation to fmid are measured. The voltage shift (ΔVtest_TG = ΔVafter_1min – ΔVinitial) corresponds to the rate of hydrolysis depending on the concentration of the TG present in the blood serum after the one minute interval.

Fig. 4.15 Calibration and
measurement modes using
C–V curves [3]

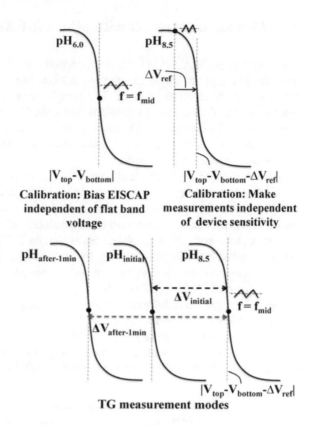

TG measurement modes

$$\Delta V_{test-TG} = 59.2\alpha(pH_{initial} - pH_{after-1min}) \qquad (4.14)$$

From Eqs. (4.12) to (4.14), Sstd_1min, S, and ΔVtest_TG, the unknown concentration of TG in the blood serum is estimated to be:

$$TG(in\ mM) = \frac{\Delta V_{test-TG}}{S \times S_{std\_1min}} \qquad (4.15)$$

The unknown TG concentration in the blood serum sample is now determined using Eq. (4.15) and the result is displayed on the LCD screen.

4.5   Measurement of TG with Mini-EISCAP

The miniaturised EISCAP TG sensors are tested using blood serum samples to esti-mate the TG content. Blood samples were taken from volunteers of both sexes in the age group from 20 to 35 years. The samples were collected at the IIT Madras Institute Hospital after obtaining ethical clearance. One part of the collected sample was tested for TG in the clinical laboratory and the other part was used in this study to validate the miniaturised EISCAP sensors. The serum samples of 10 ml are delivered to the device using a micropipette.

Figures 4.16, 4.17, 4.18 show typical C–V plots measured on miniaturised EISCAP sensors using an Agilent E4980A LCR meter. The TG measurement protocol given above is used to estimate the blood serum TG concentration. The measurement procedure is discussed taking examples of sensor D1 with blood serum sample A and sensor D2 with blood serum sample B. The pH sensitivity of sensor D1 is measured by calibrating it using the two reference pH solutions (pH 6 and pH 8.5) and is found to be 35.35 mV per pH unit. After placing the blood serum sample A in D1, the shift ΔVtest_TG (= ΔVafter_1min - ΔVinitial, indicated in Fig. 4.15) in the C–V curve from the moment the sample is placed to the end of the 1 min interval is measured to be 19.1 mV.

The inset in Fig. 4.16 clearly shows a shift in the C–V curve after 1 min to the left of the initial C–V curve confirming the hydrolysis of the TG in blood serum. From Sstd_1min, S, and the shift ΔVtest_TG (+19.1 mV), the concentration of TG in the

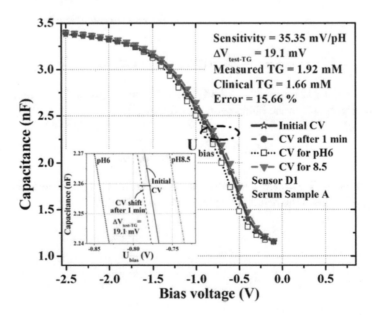

Fig. 4.16 C–V plots showing the device sensitivity and the TG hydrolysis in sensor D1 with blood serum A [3]

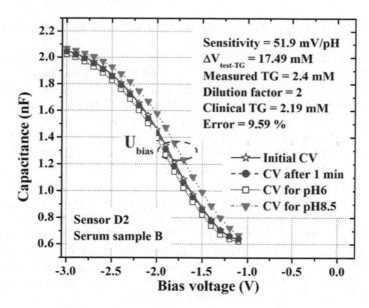

Fig. 4.17 C–V plots showing the device sensitivity and the TG hydrolysis in sensor D2 with blood serum B after diluting with Tris–HCl buffer [3]

Fig. 4.18 Enlarged plots of C–V measurements (at the mid-point of the C–V) on sensor D2 with no dilution of buffer with blood serum B [3]

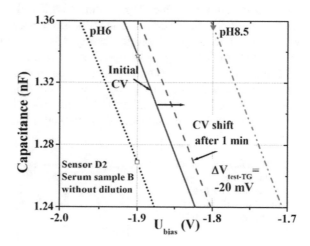

serum sample A is estimated to be 1.92 mM. The data from the measurement in the clinical lab was 1.66 mM for the same serum sample A.

Some blood serum samples, strangely, exhibited negative ΔV_{test_TG} values. These had to be diluted with Tris–HCl buffer (pH 7.4) to bring them within the normal clinical range of 0.56–1.68 mM (50–150 mg/dL, unit conversion: 1 mM $\sim$ 89 mg/dL). The dilution is necessary to reduce the effect of the buffering action of blood which is more rapid when the amount of TG in blood is higher than the normal clinical range.

A negative value of ΔVtest_TG indicates that the C–V curve after 1 min of TG hydrolysis shifts to the right of the initial C–V curve. This is a consequence of the rapid buffering action of the blood within the time interval of a minute, brought about by a high TG concentration in the sample. Measurement details for such a sample (device D2, serum sample B, Fig. 4.17) is discussed below.

The pH sensitivity of sensor D2 is measured by calibrating it using two reference pH solutions (pH 6 and pH 8.5) and it is found to be 51.9 mV per pH unit. After placing blood serum sample B in D2, the shift (ΔVtest_TG) in the C–V curve (Fig. 4.18) from the moment the sample is placed to the end of the 1 min interval results in a negative ΔVtest_TG value (20 mV), confirming the rapid blood buffering action. Figure 4.19 shows the hydrolysis of the same sample after diluting it with Tris–HCl buffer (pH 7.4) solution. The shift in the C–V curve after 1 min of TG hydrolysis is positive ΔVtest_TG (+17.49 mV), confirming that the rapid buffering action of the blood, initiated by the high TG concentration has been mitigated. Table 4.2 shows the comparison of TG concentrations in blood serum (A and B) measured using our sensors (D1 and D2) with the pathology laboratory data.

Table 4.3 compares the readout measured TG against the clinically measured TG and it gives data on the device sensitivity and C–V shift (ΔV_{test_TG}) after 1 min of hydrolysis. If the measured ΔVtest_TG is negative then the readout system indicates the need for dilution of the blood serum with Tris–HCl buffer (pH 7.4) in a 1: 1 ratio. A multiplication factor of 2 is automatically used to calculate the TG value when the diluted sample is used for testing. The error obtained from the TG readout measurements (Table 4.3) varies from 7.79 to 16.66%.

The EISCAP device sensitivity and the enzyme activity play a major role for the correct measurement of the TG concentration. The reduction in either of the sensitivities could increase the measurement error. The device sensitivity depends on the $-NH_2$ bonds on the surface of the nitride layer. But, due to the surface hydration

Fig. 4.19 Enlarged plots of C–V measurements (at the mid-point of the C–V) on sensor D2 with dilution of buffer with blood serum B [3]

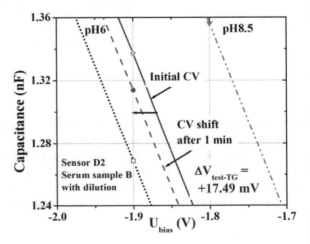

Table 4.2 Comparison of triglyceride concentrations in blood samples measured using EISCAP and a CV meter with clinical data [3]

Sensor	Serum sample	Blood serum dilution with Tris-HCI buffer pH7.4	Device sensitivity (S = V/pH)	$\Delta V_{Test\text{-}tG}$ (mV)	Measured TG (mM)	Clinical TG (mM)	% Error
D1	A	No dilution	35.35	+19.10	1.92	1.66	15.66
D2	B	No dilution	51.90	−20.00	–	2.19	–
D2	B	1:1	51.90	+17.49	2.40	2.19	9.59
D3	C	No dilution	45.60	11.00	0.86	0.77	11.86

by the electrolyte, some of the $-NH_2$ bonds get replaced by $-OH$ bonds with a lower sensitivity to pH changes. The enzyme activity could also be affected by a non-uniform surface functionalisation. The enzyme activity depends on the active site orientation on the functionalised surface and this could vary from device to device.

Table 4.4 compares the clinically measured TG concentrations with the TG concentrations measured using the readout system for the same blood samples (D and E), using two different EISCAPs ([D4, D5], [D6, D7]). The total measurement procedure, including the sample delivery, takes 5 min. A flow chart of the measurement procedure is shown in Fig. 4.20.

Table 4.5 compares the measurements using the large EISCAP and the mini-EISCAP and significant improvement can be seen in the sensitivity with the ease of measurement.

Thus, the mini-EISCAPs with the readout circuit were able to correctly detect and estimate the amount of TG present in blood serum, within the clinical range of 50 to 150 mg/dL, within a much shorter time. The sensor will also work for other bio-analytes, like urea and glucose, with appropriate enzymes and change in the calibration and measurement algorithm to suit specific applications.

Table 4.3 Comparison of TG concentrations in blood serum samples measured using the readout system with clinical data (courtesy M.S. Veeramani)

S. No	Age group	Blood serum dilution with Tris HCI buffer pH7.4	Device Sensitivity(S $= \Delta V/pH$)	$\Delta V_{test\text{-}TG}$ (mV)	Measured TG (mM)	Clinical TG (mM)	% error
1	20–25	No dilution	57.0	34.6	2.17	1.86	16.66
2	20–25	No dilution	52.0	24.0	1.65	1.46	13.01
3	20–25	No dilution	45.3	10.5	0.83	0.77	07.79
4	20–25	No dilution	53.1	09.2	0.62	0.56	10.71
5	20–25	No dilution	53.3	11.5	0.77	0.71	08.45
6	20–25	No dilution	55.7	10.1	0.65	0.60	08.33
7	20–25	1:1	49.4	18.7	2.70	2.42	11.57
8	26–30	No dilution	48.2	20.9	1.55	1.43	08.39
9	26–30	No dilution	47.0	25.3	1.92	1.66	15.66
10	26–30	No dilution	49.8	18.6	1.33	1.19	11.76
11	26–30	No dilution	52.2	26.1	1.79	1.56	14.74
12	26–30	No dilution	43.2	15.1	1.25	1.14	09.64
13	26–30	1:1	48.9	16.6	2.42	2.24	08.03
14	31–35	No dilution	51.6	31.8	2.20	1.96	12.24
15	31–35	No dilution	39.6	19.2	1.73	1.50	15.33
16	31–35	No dilution	40.2	17.5	1.55	1.39	11.51
17	31–35	No dilution	43.6	16.5	1.35	1.21	11.57
18	31–35	1:1	43.7	25.3	4.14	3.68	12.50
19	31–35	1:1	50.1	21.9	3.12	2.86	09.09

Table 4.4 Comparison of TG concentrations in the same blood samples measured clinically with those from the readout, using two different EISCAPs [3]

Sensor	Serum Sample	Blood serum dilution with TrisHCl buffer pH7.4	Device Sensitivity ($S = \Delta V/pH$)	ΔV_{test_TG} (mV)	Measured TG (mM)	Clinical TG (mM)	% error
D4	D	No dilution	50.0	17.7	1.26	1.19	06
D5	D	No dilution	49.8	18.6	1.33	1.19	12
D6	E	No dilution	40.0	14.3	1.27	1.14	11
D7	E	No dilution	43.2	15.1	1.25	1.14	09

Fig. 4.20 Flow chart of TG measurement procedure (courtesy M.S. Veeramani)

TG measurement procedure

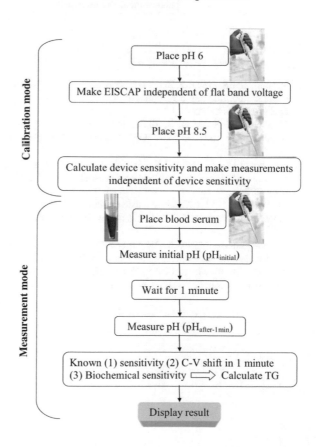

Table 4.5 Comparison of EISCAP with mini-EISCAP [3]

Parameters	Large EISCAP	Mini EISCAP
Blood serum sample volume	80 μl	5 to 10 μl
Buffer and Enzyme	Phophate buffer (pH 6), 0.25 mM; P.cepacia lipase – free enzyme	Tris–HCl buffer (pH 7.4), 200 mM; P.cepacia lipase – immobilized enzyme
Total solution volume	1.6 ml	10 μl
Device sensitivity (minimum–maximum)	30 – 55 mV/pH	35.35 – 57 mV/pH
Clinical TG reported	1.9 to 2.9 mM	0.7 to 2.86 mM
Surface area to volume ratio (mm^2 / μl)	0.0177	0.285
Measurement time	30 min + post processing time	Total readout time is within 5 min
Handling	Complex since enzyme and reference electrode are not integrated with the sensor. Post processing is required to determine the unknown TG	Simple since enzyme and reference electrode are integrated with the sensor. Readout system determines and displays the unknown TG

References

1. Veeramani MS, Shyam P, Ratchagar NP et al (2013) A miniaturized pH sensor with an embedded counter electrode and a readout circuit. IEEE Sens J 13:1941–1948. https://doi.org/10.1109/JSEN.2013.2245032
2. Vemulachedu H, Pavan S, Bhattacharya E (2008) Readout circuit design for an EISCAP biosensor. In: 2008 IEEE-BIOCAS Biomedical Circuits and Systems Conference, BIOCAS 2008, pp 73–76
3. Veeramani MS, Shyam KP, Ratchagar NP et al (2014) Miniaturised silicon biosensors for the detection of triglyceride in blood serum. Anal Methods 6:1728–1735. https://doi.org/10.1039/c3ay42274g
4. Siu WM, Cobbold RSC (1979) Basic properties of the electrolyte-SiO2-Si system: physical and theoretical aspects. IEEE Trans Elect Dev 26:1805–1815. https://doi.org/10.1109/T-ED.1979.19690
5. Garcia-Canton J, Merlos A, Baldi A (2006) A wireless potentiometric chemical sensor based on a low resistance enos capacitive structure. In: Proceedings of the IEEE International Conference on Micro Electro Mechanical Systems (MEMS), pp 462–465
6. www.cypress.com (2009) PSoC® 3. https://www.cypress.com/products/psoc-3. Accessed 1 Dec 2020

Chapter 5
Cantilever Sensors for Triglycerides and Urea

Starting with a brief genesis of cantilevers as a platform for sensing, the use of a microcantilever as a sensor for mass detection with high sensitivity is described. Though cantilevers have achieved record sensitivity for measurements in air and vacuum, measurements in liquids have limited sensitivity as well as reliability issues. Since many biological samples are present in liquid environment, we explore ways to improve these. Enzymatic hydrolysis of triglycerides and urea and consequent changes in the resonance frequency are used, though the actual working of the sensors is different for the two cases. Improvements in the immobilisation techniques, measurements at higher modes with better signal to noise ratio and improved data analysis helps to achieve both high sensitivity and reliability. Fabricating cantilevers with reduced thickness gives the lowest detection limits of 7 nM for Tributyrin (a short-chained triglyceride) and 10 nM for Urea.

Microcantilevers (MCs) are popular label free mechanical sensors, mainly due to their ability to transduce a variety of chemical and physical phenomena into mechanical movement on a μm scale [1–4]. The ability to detect interacting compounds, without the need to introduce an optically detectable label on the binding entities, is what makes cantilevers label free sensors. The present avatar of cantilevers as sensors stemmed serendipitously from the use of micromachined cantilevers as force probes in atomic force microscopy (AFM). Users found these probes, made of silicon and silicon nitride, to be extremely sensitive to a variety of environmental factors, such as acoustic noise, temperature, humidity, and ambient pressure. In 1994, two research teams, one from Oak Ridge National Laboratory and the other from IBM Zurich, developed the undesired interference in the AFM measurements with the cantilever probes into a platform for a new family of highly sensitive sensors [5, 6]. They found that a standard AFM cantilever could function as a microcalorimeter, offering fJ sensitivity.

Shifts in the resonance frequencies of microcantilevers (MCs), indicated mass-sensitivity much higher than conventional piezoelectric gravimetric sensors [7]. The sensitivity of MCs to minute quantities of adsorbates was superior to that of traditional quartz crystal microbalance (QCM) and surface acoustic wave (SAW) transducers.

© The Author(s), under exclusive license to Springer Nature Switzerland AG 2021
E. Bhattacharya, *Biosensing with Silicon*, SpringerBriefs in Materials,
https://doi.org/10.1007/978-3-030-92714-1_5

MC based sensors can also detect and identify biomolecules with orders of magnitude higher sensitivity than conventional methods such as ELISA [8] and can detect mass down to fg level, that is, even a single virus [9, 10]. The ability to detect DNA and viruses in such low concentrations can help diagnose diseases in early stages. Most of the reported high sensitivity measurements have been done in vacuum [11, 12]. Since many biological processes occur in liquid environment, it is also desirable to achieve high sensitivity for detection of biomolecules present in liquids with the MCs. The detection of triglycerides and urea in liquid environment using cantilever sensors is discussed here. Besides that, high sensitivity detection of antibody-antigen system in liquid [13] and measurement of enzyme activity [14] have also been reported earlier using resonance frequency measurements in polySi cantilevers.

5.1 Cantilevers as Sensors

MCs are simple mechanical beams supported at one end, typically 0.2–2 μm thick, 20–100 μm wide, and 100–500 μm long. The beams can also be anchored at both ends forming a bridge or a fixed–fixed beam. MCs can be routinely fabricated from various materials, including silicon, using well-established microelectronics batch processes that involve photolithographic patterning with surface and/or bulk micromachining. Process steps for a surface micromachined oxide anchored polySi cantilever beam have been discussed in Chap 2. Measurements with cantilevers can be performed in the static or the dynamic mode, depending on the application. In the static mode, the deflection of the cantilever, due to changes in the surface stress, is measured while in the dynamic mode, change in the resonance frequency of the cantilever due to additional mass loading is measured. These changes can be caused by a multitude of external stimulations like extremely small mechanical forces [15–18], charges, heat fluxes [19–21], etc. Unprecedented sensitivities have been reported with cantilevers resonating in air and vacuum, but measurements in fluids suffer from loss of sensitivity due to damping (Fig. 5.1) and the emphasis here will be on improving the measurement sensitivity in liquids.

5.2 Sensing Methods

Sensing techniques using cantilevers, for both the static and the dynamic modes, can be optical [23], piezoresistive [24], piezoelectric [25] and capacitive [7].

   Static mode: Roberto [26] showed that the bending-plate method with microfabricated cantilevers can be used to transduce the binding of a biological substance to a receptor into a signal [26]. The feasibility of the method was demonstrated by coating AFM silicon nitride MCs with the herbicide 2,4 dichlorophenoxyacetic acid and measuring the deflection, while continuously rinsing the beam with a solution containing the monoclonal antibody. The changes in the surface stress of the

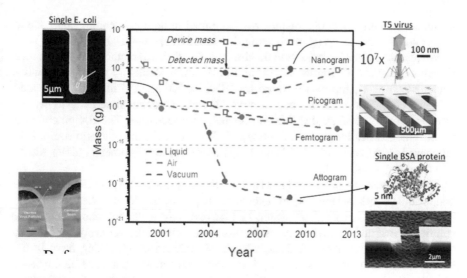

Fig. 5.1 Chronology of the measured mass by cantilevers in different media relating the device mass and the minimum detectable mass (m_{min}) [22]

cantilever due to the binding of the antibody was monitored. The beams were 190 μm long, 600 nm thick and V-shaped in order to minimize lateral deflections and a suitably modified commercial AFM was used to measure the deflections as precise as 0.1 nm. The deflection of MCs, due to surface induced stresses, caused by the enzymatic hydrolysis of glucose has been reported earlier [27]. Measurements which depend on surface induced stresses have some disadvantages [28]. The residual stresses would contribute to an initial bending of the cantilever beam resulting in erroneous readings.

Dynamic mode: Frequency-based methods for sensing are increasingly used because of their high sensitivity to adsorbed species, as compared to other methods [29, 30]. Of all the frequency-based methods like quartz crystal microbalance (QCM), piezo electric resonators, etc., MC based ones are of considerable interest because of their ability to sense small changes in physical, chemical and biological characteristics of the surrounding medium or surface [31–34]. The materials used for the fabrication of the cantilever beam resonators include polysilicon, silicon nitride as well as some of the polymers [35–37]. The silicon-based resonators have spring constant of the order 10^{-3} N m^{-1} which makes them responsive to extremely small forces of magnitude in 10^{-12} N. Hence, MCs have been extensively used for the detection of bioanalytes, measured through the change in the mass on the surface (antigen-binding) or through variation of properties (pH, viscosity, ionic strength, etc.) in the surrounding medium [38].

Although these methods have managed to achieve limit of detection as low as atto and zepto-gram, the real-time usage is limited by its operation only in air or vacuum [9, 34, 39]. For accurate real-time applications, the measurements often need to be carried out in liquid ambient, as these bioanalytes are present in soluble form in serum along with other components. Measurement in liquids is complicated due to loss of signal from damping, decreasing the sensitivity [13, 40, 41].

We restrict our measurements to the dynamic mode on (a) rectangular gold coated silicon cantilevers of lengths (l) 500 and 750 μm, a width (w) of 100 μm and a thickness (t) of 1 μm, manufactured by IBM, Zurich and (b) in-house fabricated poly-silicon cantilevers described in chap. 2, and (c) for improved sensitivity we have used in-house fabricated thin cantilevers, described later. The equipment used for the characterization were (i) Cantisens® CSR-801 research platform equipment, purchased from Concentris GmbH, and (ii) Laser Doppler vibrometer (LDV), MSA-500 from Polytec GmbH.

The Cantisens® system was used for both the frequency sweeping (FS) technique and for the continuous tracking of resonance frequency by phase locked loop (PLL) method, in DI water (deionized water, 12 MΩ resistivity), ethanol, phosphate buffer solution (PBS) and air environments. In this system the cantilever array, loaded into the measurement chamber of 5 μl volume, are excited by mounting them in the holder which contains an electrically insulated piezoactuator that is excited using a frequency generator in the system. The amplitude of the vibration of cantilever is sensed using a laser deflection system. The reflected laser from the cantilever is fed to a position sensitive detector (PSD) which generates a voltage output. This signal is compared to the excitation signal in frequency analyser, which is synchronized to the frequency generator, and generates an amplitude versus frequency plot. The frequency spectrum is generated by sweeping the frequency and plotting the amplitude of vibration at each frequency. For the PLL technique, the frequency generator is forced to operate at the resonance frequency of the cantilever. This is done by comparing the phase of the output signal with the excitation signal and then changing the excitation signal to minimize the error between the phase of the two signals. The tracking of the signal is dependent on the proportional (P) and Integral (I) gain and must be optimised. The temperature controller in the Cantisens® software was set to 25 °C, excitation voltage of 8 V and the laser intensity was set near 80% of full intensity [39].

The LDV works on the principle of Doppler shift in frequency, as the signal is reflected back by the vibrating cantilever. The detection of motion is done by a laser beam which is split into the reference and the measurement beams. The measurement beam is positioned to scan at the tip of the cantilever. The back-scattered laser beam interferes with the reference beam and the signal, detected by a photo detector is then compared with the reference signal generated by the decoder inside the LDV. The output signal is then processed using Fast Fourier transform (FFT) to produce a frequency spectrum of the vibrational velocity of the cantilever. The FS experiments were done using the LDV in DI water and air. For measurements in liquid, a fluid cell of 1.2 ml volume with a piezo actuator with wrap around electrode was fabricated, as shown in Fig. 5.2. Actuation to the cantilever is provided by mounting it on the on top of piezo-plate.

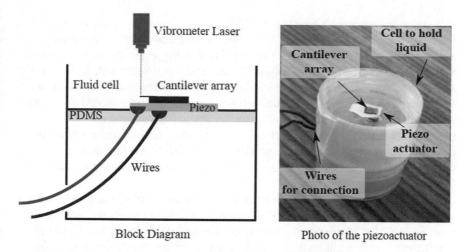

Block Diagram Photo of the piezoactuator

Fig. 5.2 Block diagram and photograph of the fluid cell with the piezoactuator for measurements with LDV [39]

5.3 Detection of Triglycerides (TG) Using Microcantilevers

Estimation of TG through the change in the pH of the electrolyte in an EISCAP, brought about by the enzymatic hydrolysis, was discussed in Chap. 3. The electrolyte consisted of KCl and PBS buffer with the sample, Tributyrin—a small chained TG—added. The buffer is required to make stable measurements but masks small changes in the pH, which limits the level of detection. The products of tributyrin hydrolysis—butyric acid and glycerol—make the solution more viscous and dense. A cantilever beam vibrating in a denser fluid has a lower resonance frequency [4] and this property can be used to detect the triglyceride concentration. We have measured the resonance frequency of a cantilever beam immersed in the solution to monitor the reaction products. The reactions are carried out either by the addition of optimized quantity of free enzymes to the triglyceride solution and immersing the beam in it. Or the beam, with the enzyme immobilised on its surface, was immersed in the triglyceride solution. We initially compare the performance of the EISCAP and MC sensors with respect to the sensitivity ranges and efficiency of these two sensors for the detection of triglycerides [42]. Though the enzyme lipase has the highest activity at pH 7, using a buffer in the electrolyte is counterproductive since it would neutralize the change in pH due to the enzymatic hydrolysis, especially at low concentrations. We found the C–V measurements became unstable at low buffer concentrations and the lowest tributyrin concentration we could measure was 1 mM. The cantilever, on the other hand, functions well even without the buffer which immediately allows measurements at much lower concentrations.

The oxide anchored polysilicon cantilever beams, of length 200 μm, width 20 μm and thickness 2 μm with 1.6 μm gap between the beam and the substrate were fabricated in house by surface micromachining [43] as described in Chap. 2. A

Doppler vibrometer, with suitable modifications, was used to detect the resonance frequency of the cantilever beams. It is a non-contact, non-destructive technique and ensures high measurement accuracy with reduced testing time. The cantilever beams were immersed in the tributyrin–lipase solution and were excited with a sine wave of 10 Vp − p coupled with a dc-offset voltage of 100 mV. The frequency of the signal was varied from 5 to 100 kHz. The frequency at which the amplitude of vibration was found to be maximum was taken as the resonance frequency of the beam. A cantilever beam immersed in water has a smaller resonance frequency since water, being denser than air, contributes an extra mass (Δm) to the cantilever mass. The resonance frequencies of the cantilever vibrating in air and water were measured as 76.4 and 48.3 kHz respectively.

Tributyrin solutions of varying concentrations were prepared in deionized water to which 1 mg lipase (1 mg/10 ml of the solution) was added and the resulting solution was incubated for 20 min for the hydrolysis. The surface micromachined polysilicon cantilever beams were immersed in the hydrolysed tributyrin solution. One molecule of tributyrin hydrolyses to one molecule of glycerol and three molecules of butyric acid.

$$\underset{\text{Tributyrin}}{C_{15}H_{26}O_6} + \underset{\text{Water}}{3H_2O} \overset{\text{Lipase}}{\rightarrow} \underset{\text{Glycerol}}{C_3H_8O_3} + \underset{\text{Butyric acid}}{3C_4H_8O_2} \tag{5.1}$$

The cantilever sensor was first calibrated using different concentrations of butyric acid and glycerol solutions, stoichiometrically equivalent to fully hydrolysed tributyrin solutions. The cantilevers were actuated in equivalent solutions of glycerol and butyric acid mixtures in which the concentrations of glycerol varied from 24–480 μM and the butyric acid concentrations were thrice that of glycerol. It was noted that the response of the sensor was linear at lower concentrations and with increasing concentrations it showed saturation in its response. The experiments were repeated using enzymatically hydrolysed tributyrin solutions and the results were found to be similar to the calibrated readings. Measurements could not be carried out for concentrations above 100 μM as bubbles started forming in the liquid which disturbed the laser spot, making the measurements unstable. We can see that the points at 150 and 250 μM have a larger variation due to this. The resonance frequency of the sensor was seen to reduce from ~48 to 28 kHz for 10–100 μM of tributyrin concentration. Figure 5.3 compares the response of the cantilever with butyric acid + glycerol solutions and tributyrin hydrolysed solutions. Due to the absence of the buffer, the cantilever sensor was able to detect tributyrin concentrations as low as 10 μM while EISCAPs did not respond to concentrations lower than 500 μM.

Fig. 5.3 Variation of resonance frequency with the concentration of enzymatically hydrolysed tributyrin (■). Also given is the calibration plot (⋆) of frequency versus concentration generated with stoichiometric mixtures of butyric acid and glycerol as discussed in the text [42]

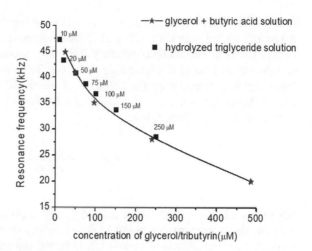

5.4 Sensitivity and Reliability for Measurements in Liquids

Analysis of dilute solutions and miniaturized devices for point of care testing demands very low-concentration ultrasensitive detection of triglycerides. Also, the pathogen antigen concentration in serum during the onset of any disease is usually present in extremely low concentration (ng ml^{-1}) which makes it essential to have a sensor with a high sensitivity in liquids for detection [44, 45]. Current methods like enzyme-linked immunosorbent assay (ELISA), radio-immunoassay show cross-reactivity and require a long time for the detection. Since the sensitivity with MCs increases with the resonance frequency, this can be achieved by fabricating a sub-micron thick cantilever beam [46, 47] and making measurements in higher modes [48].

To improve the mass sensitivity and functionality of the cantilever in liquids, several techniques and designs have been proposed such as use of in-plane modes [49] or torsional modes [50], combining spectroscopy with MC [51], embedding nanochannels inside the cantilever [52], and protecting the cantilever with a liquid–air interface [53]. Repeatability and reliability of MCs operating in an aqueous environment are important issues for bio-sensing applications. The minimum limit of detection ($m_{min} \propto m_c/Q$) depends on the mass of the cantilever (m_c) and the quality factor Q [8]. Since the signal to noise ratio (SNR) in a liquid medium is low, the dependence of m_{min} on the external factors increases [54]. Problems of repeatability and reliability have been attributed to many random uncertainties in resonance characteristics such as the actuation and measurement schemes as well as the device type. Although it is possible to reduce such uncertainties [55–57], the proposed solutions are limited in their applicability to cantilevers operated in a liquid environment.

The operation of the cantilever in an aqueous environment, for bio-sensing, poses a problem brought about by effects such as viscous damping and hydrodynamic loading, that suppress Q and reduce the sensitivity [12, 22]. Kim et al. observed, in their thiolated DNA binding measurement, that the frequency shifts were an order of magnitude higher than the number of DNA binding sites available on the cantilever surface [58]. The variation in frequency shift was attributed to hydrodynamic loading and the mechanism of DNA binding to the cantilever [8]. In liquids, the resonance frequency is dominated by the coupling of the liquid to the vibrating cantilever, thereby introducing the added mass of the liquid to be displaced along with the cantilever mass, as given below.

$$f^n_{liquid} = \frac{f^n_{vac}}{\sqrt{1 + \frac{m_{fluid}}{m_c}}} \tag{5.2}$$

where, m_{fluid} is the hydrodynamic loading and $f^n{}_{vac}$ is the frequency of the nth transverse mode in vacuum. Since the SNR is strongly dependent on the amplitude at resonance (B), amplitude of noise (A_o), Q factor and the frequency of operation of the cantilever (f) [59], this additional mass loading degrades the SNR of the MC in the liquid medium. The Q factor and the peak values can be extracted, either by fitting the resonance curve to the transfer function of a simple harmonic oscillator or to a Lorentzian curve [60]. A general expression for the amplitude function is [59],

$$S(f) = A_0 + \frac{B}{4Q^2(\frac{f}{f^n_{liquid}} - 1) + 1} \tag{5.3}$$

The quality factor of the cantilever can be improved by operating the cantilever in one of its higher modes, improving the sensitivity and the limit of detection [48]. At higher modes, the hydrodynamic mass loading and damping decrease, and damping can be neglected for mode numbers $\geqslant 3$ [61].

For the FS measurements, it was seen that the different modes show a shift in frequency in either (left or right) directions as the applied voltage to the piezo actuator was varied. The shifts in the peak values were in the range of about 500 Hz to a few kHz, for almost all the modes [39]. The reason for the frequency shift can be attributed to the increasing actuation force which in turn led to an increase in the amplitude (in picometers) of the cantilever vibration. To solve the problem of shifting resonant peak values, the amplitude (in picometers) of the vibration was fixed to be independent of the applied voltage to the piezo-actuator. Thereafter white noise actuation was able to give a stable peak value, with least deviation (deviation value in 10–20 Hz range), for different modes in the range of actuation voltages of 0.25–1.25 V range.

5.4.1 Frequency Sweep and Spurious Modes

Frequency sweep (FS) measurement was done for the cantilever array in DI water and ethanol, without flow, using the Cantisens®. The amplitude response of the resonator was measured with an active drive signal close to its resonance frequency. The frequency generator sweeps the frequency every 100 ms (ten samples every second) with a step size of 50 Hz, i.e. a sweep of 500 Hz every second. The parameters were kept constant for each sweep frequency range from 10 to 1000 kHz. For each cantilever, several frequency spectra were taken over a duration of 8–12 h per day (different intervals), repeated over a few days (different runs). For the LDV, the frequency sweep was done in air, and in DI water, and then the frequency spectra were compared with the Cantisens® data. Two types of signals were used for actuation (i) white noise (a random signal with flat frequency spectrum) and (ii) periodic chirp (signal to excite all the FFT lines of the measured spectrum, amplitude being same for all the frequencies) [Polytec GmbH Polytec Scanning Vibrometer Theory Manual] signal with a resolution of 19.8 Hz. An excitation voltage of 1.0 V for white noise and 0.75 V for periodic chirp was fixed for the experiments in air. For the measurements in liquid, only periodic chirp was used for the actuation with an excitation voltage of 5 V, as with white noise signal the Q values were found to be low [39].

Figure 5.4 shows the FS measurements in DI water and ethanol from the Cantisens® from which the vibration modes have to be identified correctly for sensing. This plot shows more peaks than expected and it is important to eliminate the spurious modes. Of the different mode shapes of the cantilever vibration

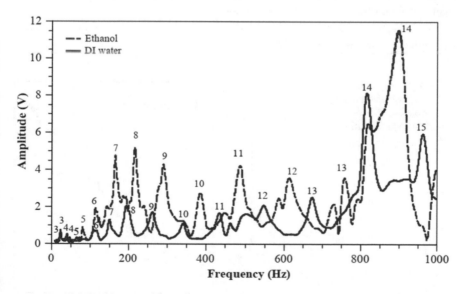

Fig. 5.4 Frequency spectrum of cantilever in DI water and ethanol with the regular modes identified [39]

namely transverse, longitudinal, torsional and lateral, generally the transverse mode is preferred for mass detection. If the resonance frequency of the other type of modes are nearby then those can interfere, leading to spurious modes in the spectrum (Fig. 5.4). Other sources of spurious modes are: (i) spot size or position covering several nodes, crests and troughs of the cantilever vibration (ii) coupling between the cantilever and the mounting holder [62, 63] and (iii) mode coupling. As the linewidths of different modes increases, because of low SNR, they can couple resulting in merging or additional peaks. When the quality factor is high, these modes can be distinguished easily [64, 65].

The regular modes can be segregated from the spurious modes by plotting $f_n - f_{n-1}$ versus n as shown in Fig. 5.5. The difference of the modes follows a monotonic increasing function. When the frequency difference is plotted against the mode number (n), spurious modes will not follow the monotonic increase and can be easily identified.

The exact peak values has to be determined to measure any shifts in the resonant frequencies. The SNR in a liquid medium is low, and the resonance peaks also showed drift and noise. Therefore, m_{min} of the system (which was previously $\propto m_c/Q$) is now decided by the accuracy with which we can determine the peak value of that particular mode. To estimate the error in the peak value, the standard deviation (SD) was calculated by taking ten frequency spectra over a time period (done on two different cantilever array, four different cantilevers in each) and then finding the mean value and standard deviation of each of the resonant frequencies. For example, at the 14th mode the peak value was 814.99 ± 0.557 kHz, and therefore the minimum frequency shift that can be accurately measured at the 14th mode should be greater than 0.557 kHz for a reliable mass determination.

The LDV is a stable system with a smaller spot size (Fig. 5.6) and the FS spectra does not show any spurious modes.

Fig. 5.5 $f_n - f_{n-1}$ *versus n* plot for the spectrum in DI water from Fig. 5.4 to identify the spurious modes [39]

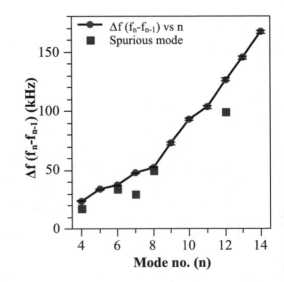

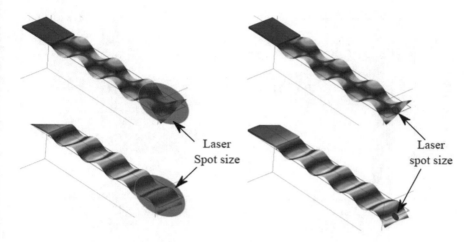

Fig. 5.6 Schematic comparison of spot size on the cantilever in Cantisens® and the LDV [39]

Calculation of mass from the FS improves in detection accuracy when multiple resonant peaks are used. The peak values of all the resonant peaks were extracted by a Lorentzian fit. Averaging of these peak values, accurately resolved the minimum measurable frequency shift for a particular mode, and the m_{min} was found to be greater than 1 ng for all modes.

5.4.2 PLL Tracking

For PLL tracking in Cantisens® the 14th resonant mode was chosen due to its high amplitude, as shown in Fig. 5.4. A stable baseline is necessary for sub-nanogram measurements and appropriate values of the proportional (P) and integral (I) gains have to be chosen to achieve this. The values P = 350 and I = 800 provided the most satisfactory stable baseline in DI water environment, in no flow and with flow conditions, for extended time period as shown in Fig. 5.7. An initial time period is required for the resonant frequency to stabilize after which the DI water intake is done at 4 μl s^{-1}. Post uptake, the PLL tracking was carefully monitored for the baseline to return back to its initial value. The resonance frequency value can drift because of the heating by the laser [66–68], as the laser spot size is large at 250 μm, but even after tracking for a very long time (20–30 min), no drift was seen in the resonance frequency values. In DI water, a minimum error of 5 Hz was measured at the 14th mode leading to m_{min} value of 14 pg. PLL tracking was found to be much more sensitive and reliable than FS, if the baseline is stable, which is difficult to obtain in liquids. A stable baseline was achieved by optimizing for different proportional (P) & integral (I) gain values. At lower values, tracking was slow with PLL locking out, while higher values with faster tracking amplified the error. Optimized values

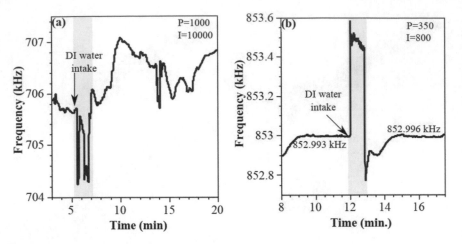

Fig. 5.7 PLL tracking with different values of proportional (*P*) and integral (*I*) gain **a** with unstable baseline and **b** stable baseline in DI water [39]

of P $=$ 350 and I $=$ 800 gave the least error of 5 Hz at 14th mode leading to a m_{min} value of 14 pg, which indicates an improved limit of detection when compared to the FS technique. Using the higher mode, detection of Triglyceride at **50 µM** level of concentration was achieved.

5.5   Improving the Sensitivity

Reduced thickness of the beams, in conjunction with appropriate resonance modes, can also suitably mitigate the damping and hydrodynamic loading issues [69], and improve the sensitivity of detection of biomolecules in liquids to a great extent. Towards this, polysilicon beams of 165 nm thickness were fabricated [46, 47] to enable low-concentration detection of bioanalytes in liquids. Polysilicon was chosen because of its high Q-factor attributed to high density and acoustic velocity of the material [70].

5.5.1   Fabrication of Thin Polysilicon Cantilevers

A 165 nm thick undoped polysilicon layer was deposited by low-pressure chemical vapour deposition of silane at 620 °C on a sacrificial layer of 1 µm oxide, thermally grown on a (100) double side polished p-type silicon wafer. The polysilicon layer was annealed at 950 °C for half an hour, after the deposition, to remove the residual stress. A double-sided mask aligner (MA6/BA6) was used for patterning. The polysilicon beam was etched from the top by reactive ion etching (RIE) using an ICP-RIE system and SF_6 chemistry. Patterning followed by etching the substrate silicon from below was done using the Bosch process in a deep reactive ion etching system. The beam was finally released by removing the sacrificial oxide with hydrofluoric acid (HF) vapour etching for 5 min [13]. The schematic of the process steps is given below (Fig. 5.8).

The fabricated polysilicon cantilever beam with a thickness of 165 nm, 5.12 μm width and 53 μm length are shown in Fig. 5.9. The dimensions were optimised for the mass to be in pico-gram range and $f_{max} < 6$ MHz, the maximum limit of the LDV [71]. The l/w ratio has to be ≥ 4 to reduce the effect of damping and the hydrodynamic loading at higher modes [61]. The sensitivity of the bare cantilever was tested by measuring the mass of a gold layer deposited on the beam. The cantilever showed a high sensitivity of 280.4 fg kHz^{-1} in liquids which was higher than most of the mass sensitive mechanical devices reported previously (Table 5.1).

The Young's modulus (E) for the cantilever was calculated using the formula (Digilov and Abramovich [83]:

$$E = \left(f_R \times (3\rho_c)^{0.5} \times \frac{4\pi}{t} \times \frac{l^2}{c^2} \right)^2 \tag{5.4}$$

where f_R is the measured resonant frequency at 1st mode in air for the cantilever (75.203 kHz), ρ_c is the density of polysilicon (2330 kg m^{-3}), t and l are the thickness (165 nm) and the length (53 μm) of the beam and c is the eigen constant (1.875). The Young's modulus for polysilicon was calculated to be 144 GPa. This was later used to determine the hydrodynamic function and subsequent determination of attached mass.

Following the protocol described earlier for the immobilization of lipase, the polysilicon beam (and a 1 × 1 cm^2 piece of polysilicon) were first exposed to an O_2 plasma for 15 min followed by immersion in toluene containing 1% APTES at 80 °C for 1 h in a soxhlet apparatus. This was followed by washing in toluene, ethanol, deionised water and curing at 120 °C for 1 h. Samples were then immersed in a solution containing 0.5% glutaraldehyde in DI water for 30 min at room temperature. Finally, the samples were incubated in the lipase enzyme solution with the concentration previously optimised for silicon wafer (10.31 ± 1.1 μg cm^{-2}) at 4 °C overnight. The immobilized protein concentration was obtained by bicinchoninic acid (BCA) assay using known lipase concentration as standard. The activity of the immobilized lipase was estimated by a para-nitrophenyl butyrate (pNPB) assay. The hydrolysis of pNPB was measured for a fixed time using a UV–Vis Spectrophotometer at 415 nm. The BCA assay showed that the saturating concentration for the enzyme lipase on the polysilicon layer was 10.31 ± 1.1 μg cm^{-2} with an activity of 2.7 units cm^{-2} measured by the pNPB assay. The K_m of the immobilized lipase was higher (0.843 mM) due to the restricted orientation of catalytic site compared to the free lipase enzyme which showed a K_m of 0.343 mM (Fig. 5.10) [84]. The catalytic efficiency (k_{cat}/K_m) for the free lipase was much higher (19 M^{-1} min^{-1}) as compared to the immobilized lipase (2.75 M^{-1} min^{-1}).

Fig. 5.8 Process flow for the fabrication of the thin polysilicon cantilevers [13]

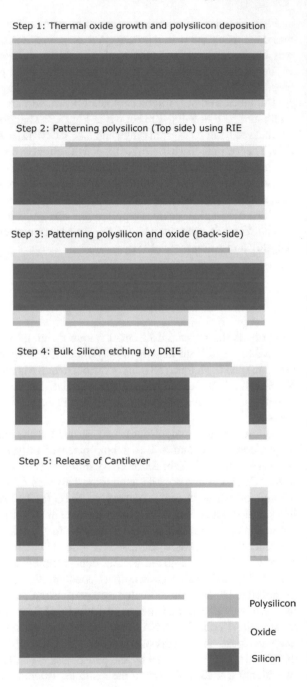

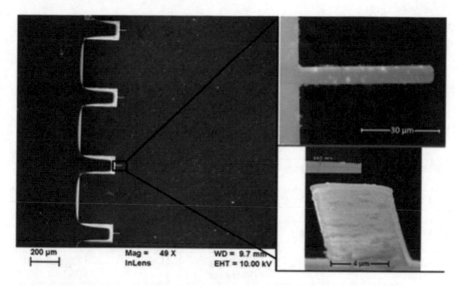

Fig. 5.9 Fabricated polysilicon cantilever. Inset: magnified image of the beam immobilised with the enzyme [13]

5.6 Detection of Triglycerides Using LDV

The detection of triglycerides was carried out by measuring the change in frequency at the 8th mode, after the lipase mediated hydrolysis of tributyrin in deionised water. The specific gravity of tributyrin is 1.035 g cm^{-3} which is closer to the aqueous medium. On hydrolysis, glycerol (1.26 g cm^{-3}) and butyric acid (0.952 g cm^{-3}) are formed, thus increasing the density of the medium [85]. For calibration, the measurement was also performed using standard solutions of glycerol and butyric acid (1:3), which are the products of the hydrolysis reaction (Fig. 5.11).

The system was tested using both free lipase in the medium and with lipase immobilized on the polysilicon cantilever. On immobilization, the frequency of the device was found to decrease by 54 Hz due to the mass of lipase attached. Further, both the systems were found to be responsive to the tributyrin concentration in the range of 7 nM to 500 nM (Fig. 5.12). From 100–250 nM a saturation is observed in the response of cantilever frequency to the tributyrin hydrolysis and is possibly due to the limited quantity of enzyme available. The sensitivity of the free lipase system was found to be lower with Δf of 410 kHz mM^{-1} of tributyrin as compared to the lipase immobilized on the polysilicon beams with Δf of 522 kHz mM^{-1}. This could be because of the close proximity of the glycerol formed to the functionalised beam (Fig. 5.12: inset).

The smaller cantilever could achieve a limit of detection of 7 nM, which is much lower than the previously reported value using silicon resonator [42]. The high signal to noise ratio, makes the sensor viable when compared to amperometric sensors which usually need extensive modification and immobilization of multienzymes for

Table 5.1 Reported non-specific mass sensing using Micro/Nanomechanical devices [13]

Type of the sensor	Limit of detection	Sensitivity	Measurement medium	Reference
Microelectromechanical disk resonators	–	15 pg Hz^{-1}	Liquid	Mehdizadeh et al. [72]
Resonating microplates	20 ng	–	Liquid	Mahajne et al. [73]
Microchannel resonators	80 ag	15 ag Hz^{-1}	Liquid	Arlett and Roukes [74]
NEMS resonator	2.53 ag	1 ag Hz^{-1}	Air	Ekinci et al. [75]
Cantilever array	0.86 ng	284 ng Hz^{-1}	Liquid	Ghatkesar et al. [76]
Nano-optomechanical disk	5 pg	–	Liquid	Gil-Santos et al. [77]
Micromechanical resonators	8.3 pg	27 ppm ng^{-1} (37 µg Hz^{-1})	Liquid	Tappura et al. [78]
Ring Resonator	1 ng	–	Air	Fukuyama et al. [79]
Microresonator	9.5 fg	0.16 ag Hz^{-1}	Air	Gupta et al. [9]
Single-crystal silicon micro-oscillator	8 pg	0.35 pg Hz^{-1}	Air	Zhang and Turner [80]
Cantilever	10 pg	–	Air	Spletzer et al. [81]
Microcantilever	450 ng	122 pg Hz^{-1}	Air	Faegh et al. [82]
Nanoresonator	–	**280.4 fg kHz^{-1}**	**Liquid**	Chinnamani et al. [13]

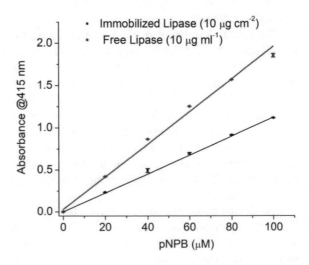

Fig. 5.10 pNPB assay for the estimation of lipase activity [13]

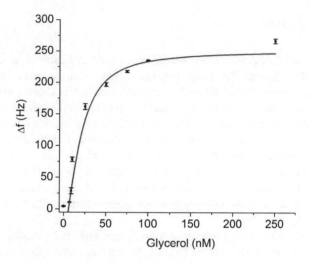

Fig. 5.11 LDV measurement using various concentration of glycerol [13]

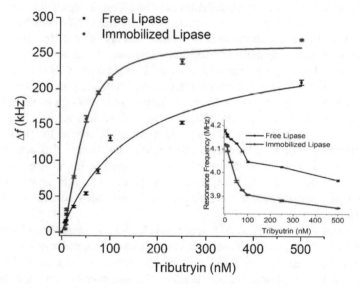

Fig. 5.12 Change in resonance frequency with Tributyrin concentration. Inset: resonance frequency versus tributyrin concentration [13]

coupled reactions [86–88]. Besides being useful for small blood volumes, the range of detection is also suitable for samples with trace amounts of TG like peritoneal, pleural, synovial, spinal fluids, urine, etc. Presence of TG and glycerol in these fluids are effective markers for various diseases [89–92].

5.7 Urea Sensor

Frequency-based methods for the detection of urea are limited in number. The commonly used devices for frequency-based detection includes quartz crystal microbalance, piezo-electric resonator, surface acoustic resonator etc. and these methods rely upon the measurement of change in pH or conductivity caused during the enzymatic reaction with urease. We have used a novel method of using aluminium coated cantilevers [93]. Ammonium ions have been effectively used for etching of metals [94]. Ammonium hydroxide, in the presence of metal, forms metal hydroxide resulting in its dissolution. Grubisic et al. in 2009 reported selective etching of aluminium using ammonia vapour [95]. Previous reports have used aluminium metal in ammonium hydroxide as a strategy for the gradual introduction of aluminium as dopant into thin films [96].

We utilise the Al etching property of ammonium hydroxide, produced during the hydrolysis reaction in the presence of urease, for the measurement of the urea concentration [93]. Al films were deposited on the polysilicon cantilevers and functionalized with the enzyme urease. The ammonium ion produced during the hydrolysis of urea is expected to etch the aluminium film. The resulting change in the mass, can be measured as a shift in the resonance frequency of the cantilever. To validate the result, direct etching of Al using ammonium hydroxide was also studied for comparison.

On the polysilicon cantilevers of 165 nm thickness, 53 μm length and 5.12 μm width, described above, aluminium film of thickness 50 nm was deposited using e-beam evaporation. The polysilicon surface, coated with aluminium, was silanised by first exposing to O_2 plasma for 15 min followed by immersion in 1% APTES in toluene at 80 °C for 1 h. Samples were cured at 120 °C and urease (10 μg ml^{-1}) was finally immobilized using 0.5% glutaraldehyde at 4 °C overnight. The unreacted sites of glutaraldehyde were blocked using 1 mM ethanolamine. The activity of the urease immobilized on Al-polySi wafer was determined by measuring the quantity of ammonia produced during the reaction. The concentration of urease on the Al coated polySi wafer, measured using bicinchoninic acid assay, was found to be 9.92 ± 0.86 μg cm^{-2}.

EDAX analysis revealed that Al was present at a total concentration of 1 atomic% to the silicon concentration of 65 atomic% on the beam. As the aluminium is present in very low concentration, it is likely to be affected by even a small amount of etchant present in the solution, thus increasing the sensitivity of the nanoresonator. The resonance frequency was measured using the Laser Doppler vibrometer (LDV) model MSA500 by frequency-sweep method. The change in mass due to Al-etching was determined from the measured change in the frequency.

Urease immobilized on Al coated polySi layer was used to study the enzymatic breakdown of urea. Etching is commonly used for smoothening and patterning of metal film. It was observed that with the increasing concentration of urea, the smoothness of the Al-film improved due to the ammonium hydroxide produced during the reaction. The average surface roughness of the Al-film measured using surface profiler decreased from 2.57 ± 0.1 to 1.5 ± 0.007 nm (Fig. 5.13).

Fig. 5.13 Surface roughness of Urease-coated polySi layer with increasing urea concentration [93]

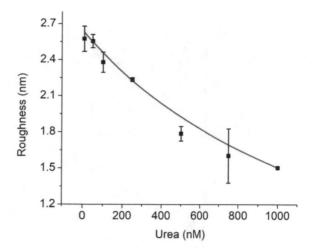

On deposition of Al, there was a decrease of 88.4 ± 6 kHz in the resonance frequency of the polySi cantilever beam, corresponding to a mass of 26.55 ± 2 picogram of Al followed by a further decrease in frequency by 9.5 ± 1.8 kHz after the immobilization of urease, due to the mass of the protein and APTES. On addition of urea, the urease enzyme immobilised on the beam, hydrolyses it to ammonium hydroxide and carbonic acid. The ammonium hydroxide produced etches the Al by the formation of aluminium hydroxide [96]. This results in the loss of mass from the nanoresonator which contributes to an increase in the resonance frequency. The change in resonance frequency was found to be directly proportional to the urea concentration with a linear response in the range 10–1000 nM of urea (Fig. 5.14). The corresponding change in the mass due to the etching of Al from the resonator surface varied from 0.525 ± 0.07 pg to 2.6 ± 0.26 pg (Fig. 5.14, Inset). The sensitivity of the device was 6.3 Hz nM^{-1} of urea with a limit of detection (LoD) of 10 nM (Fig. 5.14). Previously reported work, with frequency-based sensor, using Y-type ring resonator obtained a sensitivity of 2.7 Hz mM^{-1}. They have detected urea based on the exothermic nature of hydrolysis, therefore the change in frequency is driven by calorimetry, which is prone to interference caused by change in the temperature [97]. In another study, QCM was utilized along with fullerene to adsorb NH$_4^+$ formed during the urease reaction [98]. This gave a sensitivity of about 80 Hz per decade of the urea concentration in the range of 10^{-1} to 10^{-4} M [98]. Fernandez et al. in [14] used the volatility of ammonia produced during the urease reaction to quantify the urea concentration. In this study there was also simultaneous evaporation of water which resulted in interference to actual urea measurements [14]. In comparison to the above-mentioned methods, the present approach is more reliable and sensitive for the quantification of urea. To extend the applicability of this device as an ammonium ion sensor, calibration experiments were performed for different concentrations of ammonium hydroxide. It was observed that the device was linear for 10–1000 nM of ammonium hydroxide with a LoD of 10 nM (Fig. 5.15).

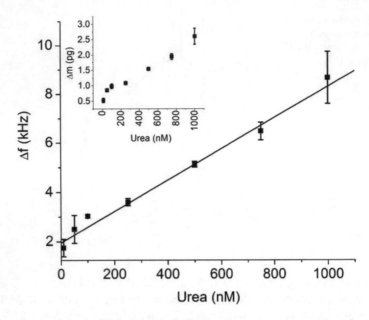

Fig. 5.14 Change in resonance frequency of Al-pSi with varying concentration of urea. Inset: Change in mass (Δm) versus concentration of urea [93]

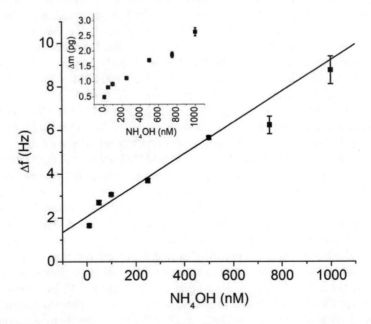

Fig. 5.15 Change in resonance frequency of Al-pSi with varying concentration of ammonium hydroxide. Inset: Change in mass (Δm) versus concentration of ammonium [93]

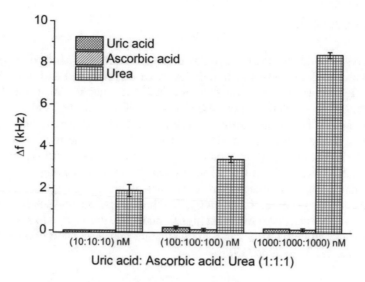

Fig. 5.16 Effect of interferents on the sensitivity of Al-pSi nanoresonator [93]

Since, samples like blood serum has a variety of biomolecules, the sensitivity of the nanoresonator was also tested in presence of interferents like ascorbic acid and uric acid. Both the interferents were tested individually and in the presence of urea. It can be clearly observed that uric acid and ascorbic acid, in the absence of urea, had negligible effect on the performance of the nanoresonator (Fig. 5.16). Thus, it shows that the nanoresonator immobilized with urease is very specific and sensitive for urea and is not affected by interference from other biomolecules in the sample.

Previously, very few groups have reported on urea sensors with such a low limit of detection. Dutta et al. in [99] devised a graphene oxide based electrochemical sensor for urea with of LoD 11.7 fM [99]. But these sensors are non-enzymatic, thereby increasing the chances of interference in presence of other non-specific redox molecules. Normal range for urea in serum is usually within 2.5–8.6 mM. But in other body fluids like cerebro spinal fluid CSF, concentration as low as 66 μM has been reported and is one of the widely used biomarkers for Alzheimer [100]. Advantage of low limit of detection allows the use of diluted samples thus reducing the effects of interferents in the sample.

In addition to the body fluids, urea and ammonia is also an indicator environmental pollution as it is one of the major ingredients found in fertilizer. The contamination in the water bodies are in a very small scale (parts per billion-ppb) and hence difficult to detect. Urea contamination in lakes has been reported to be as low as 0.25 μM and is one of the factors contributing to algal blooms [101]. Similarly, ammonia contamination varies from 55–350 μM in the water bodies [102]. The nanoresonators fabricated in this study can detect very low concentrations of urea and ammonia, thus making it suitable for wide range of applications.

5.8 Conclusion

Improved sensitivity of measurements with the resonance frequency of polysilicon and silicon cantilever beams in liquids has been demonstrated and used for estimating triglyceride and urea concentration in liquid samples. We have chosen a three-pronged approach to improve the sensitivity of measurements in liquid. First, was to improve the functionalization protocol ensuring uniformity and optimum attachment to the beam surface. Second, was to improve the measurements and data analysis, like choice of the modal frequency, rejection of spurious modes, etc. [39]. And third, finally, the culmination of using the above two along with beams of reduced thickness, and not the other dimensions, to present enough surface for the immobilization to get the high sensitivity. The thickness of the polysilicon cantilever beam was 165 nm and measurement at higher modes like the 9th and 14th contributed to the high sensitivity. Use of the same technique improved the detection limit of antigen mass to 434 ± 59 fg [13].

References

1. Basu AK, Basu A, Bhattacharya S (2020) Micro/Nano fabricated cantilever based biosensor platform: A review and recent progress. Enzyme Microb Technol 139. https://doi.org/10.1016/j.enzmictec.2020.109558
2. Moulin AM, O'Shea SJ, Welland ME (2000) Microcantilever-based biosensors. Ultramicroscopy 82:23–31. https://doi.org/10.1016/S0304-3991(99)00145-X
3. Raiteri R, Grattarola M, Butt HJ, Skládal P (2001) Micromechanical cantilever-based biosensors. Sensors Actuators, B Chem 79:115–126. https://doi.org/10.1016/S0925-4005(01)00856-5
4. Tamayo J, Humphris ADL, Malloy AM, Miles MJ (2001) Chemical sensors and biosensors in liquid environment based on microcantilevers with amplified quality factor. In: Ultramicroscopy. North-Holland, pp 167–173
5. Barnes JR, Stephenson RJ, Welland ME et al (1994) Photothermal spectroscopy with femtojoule sensitivity using a micromechanical device. Nature 372:79–81. https://doi.org/10.1038/372079a0
6. Thundat T, Warmack RJ, Allison DP, Jacobson KB (1994) Critical Point Mounting of Kinetoplast DNA for Atomic Force Microscopy. Scanning Microsc 8:23–30
7. Wachter EA, Thundat T (1995) Micromechanical sensors for chemical and physical measurements. Rev Sci Instrum 66:3662–3667. https://doi.org/10.1063/1.1145484
8. Johnson BN, Mutharasan R (2012) Biosensing using dynamic-mode cantilever sensors: A review. Biosens Bioelectron 32:1–18. https://doi.org/10.1016/j.bios.2011.10.054
9. Gupta A, Akin D, Bashir R (2004) Single virus particle mass detection using microresonators with nanoscale thickness. Appl Phys Lett 84:1976–1978. https://doi.org/10.1063/1.1667011
10. Ilic B, Yang Y, Craighead HG (2004) Virus detection using nanoelectromechanical devices. Appl Phys Lett 85:2604–2606. https://doi.org/10.1063/1.1794378
11. Buchapudi KR, Huang X, Yang X et al (2011) Microcantilever biosensors for chemicals and bioorganisms. Analyst 136:1539–1556. https://doi.org/10.1039/c0an01007c
12. Sader JE, Larson I, Mulvaney P, White LR (1995) Method for the calibration of atomic force microscope cantilevers. Rev Sci Instrum 66:3789–3798. https://doi.org/10.1063/1.1145439

13. Chinnamani MV, Bhadra P, Fidal VT et al (2020) Ultrasensitive detection of antigen–antibody interaction and triglycerides in liquid ambient using polysilicon cantilevers. J Micromechanics Microengineering 30. https://doi.org/10.1088/1361-6439/abb992

14. Fernandez RE, Bhattacharya E, Chadha A (2010) Dynamic response of polysilicon microcantilevers to enzymatic hydrolysis of urea. Int J Adv Eng Sci Appl Math 2:17–22. https://doi.org/10.1007/s12572-010-0007-6

15. Butt HJü, (1996) A sensitive method to measure changes in the surface stress of solids. J Colloid Interface Sci 180:251–260. https://doi.org/10.1006/jcis.1996.0297

16. Samuel J, Brinker CJ, Frink LJD, van Swol F (1998) Direct measurement of solvation forces in complex microporous media: A new characterization tool. Langmuir 14:2602–2605. https://doi.org/10.1021/la980073k

17. Frink LJD, Van Swol F (2000) A common theoretical basis for surface forces apparatus, osmotic stress, and beam bending measurements of surface forces. Colloids Surfaces A Physicochem Eng Asp 162:25–36. https://doi.org/10.1016/S0927-7757(99)00253-8

18. Green CP, Sader JE (2002) Torsional frequency response of cantilever beams immersed in viscous fluids with applications to the atomic force microscope. J Appl Phys 92:6262–6274. https://doi.org/10.1063/1.1512318

19. Rugar D, Yannoni CS, Sidles JA (1992) Mechanical detection of magnetic resonance. Nature 360:563–566. https://doi.org/10.1038/360563a0

20. Raiteri R, Butt HJ, Grattarola M (2000) Changes in surface stress at the liquid/solid interface measured with a microcantilever. Electrochim Acta 46:157–163. https://doi.org/10.1016/S0013-4686(00)00569-7

21. Mamin HJ, Rugar D (2001) Sub-attonewton force detection at millikelvin temperatures. Appl Phys Lett 79:3358–3360. https://doi.org/10.1063/1.1418256

22. Tamayo J, Kosaka PM, Ruz JJ et al (2013) Biosensors based on nanomechanical systems. Chem Soc Rev 42:1287–1311. https://doi.org/10.1039/c2cs35293a

23. Tang T, Xu B, Welch J, Castracane J (2004) Cantilever-based mass sensor for immunodetection of multiple bioactive targets. In: Microfluidics, BioMEMS, and Medical Microsystems II. SPIE, p 89

24. Gotszalk T, Grabiec P, Rangelow IW (2000) Piezoresistive sensors for scanning probe microscopy. Ultramicroscopy 82:39–48. https://doi.org/10.1016/S0304-3991(99)00171-0

25. Chu J (1997) Novel high vacuum scanning force microscope using a piezoelectric cantilever and the phase detection method. J Vac Sci Technol B Microelectron Nanom Struct 15:1551. https://doi.org/10.1116/1.589398

26. Raiteri R, Nelles G, Butt HJ et al (1999) Sensing of biological substances based on the bending of microfabricated cantilevers. Sensors Actuators, B Chem. 61:213–217

27. Subramanian A, Oden PI, Kennel SJ et al (2002) Glucose biosensing using an enzyme-coated microcantilever. Appl Phys Lett 81:385–387. https://doi.org/10.1063/1.1492308

28. Lavrik NV, Sepaniak MJ, Datskos PG (2004) Cantilever transducers as a platform for chemical and biological sensors. Rev Sci Instrum 75:2229–2253. https://doi.org/10.1063/1.1763252

29. Kim S, Ono T, Esashi M et al (2006) Capacitive resonant mass sensor with frequency demodulation detection based on resonant circuit Capacitive resonant mass sensor with frequency demodulation detection. Appl Phys Lett 053116:1–4. https://doi.org/10.1063/1.2171650

30. Nguyen V-N, Baguet S, Lamarque C-H, Dufour R (2015) Bifurcation-based micro- / nanoelectromechanical mass. Nonlinear Dyn 79:647–662. https://doi.org/10.1007/s11071-014-1692-7

31. Gonzalo G-M, Enrique Alonso B, Hubert P et al (2011) Development of a Mass Sensitive Quartz Crystal Microbalance (QCM)-Based DNA Biosensor Using a 50 MHz Electronic Oscillator Circuit. Sensors 11:7656–7664. https://doi.org/10.3390/s110807656

32. Li X, Tang G, Shen Z, Yong K (2015) Resonance frequency and mass identification of zeptogram-scale nanosensor based on the nonlocal beam theory. Ultrasonics 55:75–84. https://doi.org/10.1016/j.ultras.2014.08.002

33. Wang S, Shan X, Patel U et al (2010) Label-free imaging, detection, and mass measurement of single viruses by surface plasmon resonance. Proc Natl Acad Sci U S A 107:16028–16032. https://doi.org/10.1073/pnas.1005264107

34. Yang YT, Callegari C, Feng XL et al (2006) Zeptogram-Scale Nanomechanical Mass Sensing. Nano Lett 6:583–586
35. Datar R, Kim S, Jeon S et al (2009) Cantilever Sensors : Nanomechanical Tools for Diagnostics. MRS Bull 34:449–454
36. Patkar RS, Kandpal M, Gilda N et al (2014) Polymer-Based Micro / Nano Cantilever Electro-Mechanical Sensor Systems for Bio / Chemical Sensing Applications. Micro Smart Devices Syst 403–422. https://doi.org/10.1007/978-81-322-1913-2
37. Verd J, Teva J, Abadal G et al (2006) System on chip mass sensor based on polysilicon cantilevers arrays for multiple detection. Sensors Actuators A Phys 132:154–164. https://doi.org/10.1016/j.sna.2006.04.002
38. Weigert S, Dreier M, Hegner M (1996) Frequency shifts of cantilevers vibrating in various media. Appl Phys Lett. https://doi.org/101063/111733469:2834-2836.doi:10.1063/1.117334
39. Kathel G, Shajahan MS, Bhadra P et al (2016) Measurement and reliability issues in resonant mode cantilever for bio-sensing application in fluid medium. J Micromechanics Microengineering 26. https://doi.org/10.1088/0960-1317/26/9/095007
40. Riesch C, Reichel EK, Keplinger F, Jakoby B (2008) Characterizing Vibrating Cantilevers for Liquid Viscosity and Density Sensing. J Sensors 2008. https://doi.org/10.1155/2008/697062
41. Gologanu M, Bostan CG, Avramescu V, Buiu O (2012) Damping effects in MEMS resonators. Proc Int Semicond Conf CAS 1:67–76. https://doi.org/10.1109/SMICND.2012.6400695
42. Fernandez RE, Hareesh V, Bhattacharya E, Chadha A (2009) Comparison of a potentiometric and a micromechanical triglyceride biosensor. Biosens Bioelectron 24:1276–1280. https://doi.org/10.1016/j.bios.2008.07.054
43. Bhat S, Bhattacharya E (2007) Parameter extraction from simple electrical measurements on surface micromachined cantilevers. J Micro/nanolithography, MEMS, MOEMS 6. https://doi.org/10.1117/1.2794291
44. Skjøt RL, Oettinger T, Rosenkrands I et al (2000) Comparative evaluation of low-molecular-mass proteins from Mycobacterium tuberculosis identifies members of the ESAT-6 family as immunodominant T-cell antigens. Infect Immun 68:214–220
45. Sobanski MA, Barnes RA, Gray SJ et al (2000) Measurement of serum antigen concentration by ultrasound-enhanced immunoassay and correlation with clinical outcome in meningococcal disease. Eur J Clin Microbiol Infect Dis 19:260–266
46. Weeks BL, Camarero J, Noy A et al (2006) A microcantilever-based pathogen detector. Scanning 25:297–299. https://doi.org/10.1002/sca.4950250605
47. Campbell GA, Mutharasan R (2008) Near real-time detection of Cryptosporidium parvum oocyst by millimeter-sized cantilever biosensor. Biosens Bioelectron 23:1039–1045. https://doi.org/10.1016/j.bios.2007.10.017
48. Ghatkesar MK, Barwich V, Braun T et al (2007) Higher modes of vibration increase mass sensitivity in nanomechanical microcantilevers. Nanotechnology 18. https://doi.org/10.1088/0957-4484/18/44/445502
49. Tao Y, Li X, Xu T et al (2011) Resonant cantilever sensors operated in a high-Q in-plane mode for real-time bio/chemical detection in liquids. Sensors Actuators, B Chem 157:606–614. https://doi.org/10.1016/j.snb.2011.05.030
50. Hui X, Vitard J, Haliyo S, Régnier S (2007) Enhanced sensitivity of mass detection using the first torsional mode of microcantilevers. In: Proceedings of the 2007 IEEE International Conference on Mechatronics and Automation, ICMA 2007. pp 39–44
51. Finot E, Rouger V, Markey L et al (2012) Visible photothermal deflection spectroscopy using microcantilevers. Sensors Actuators, B Chem 169:222–228. https://doi.org/10.1016/j.snb.2012.04.072
52. Lee J, Shen W, Payer K et al (2010) Toward attogram mass measurements in solution with suspended nanochannel resonators. Nano Lett 10:2537–2542. https://doi.org/10.1021/nl101107u
53. Yu H, Chen Y, Xu P et al (2016) μ-'Diving suit' for liquid-phase high-Q resonant detection. Lab Chip 16:902–910. https://doi.org/10.1039/c5lc01187f

54. Sader JE, Sanelli J, Hughes BD et al (2011) Distortion in the thermal noise spectrum and quality factor of nanomechanical devices due to finite frequency resolution with applications to the atomic force microscope. Rev Sci Instrum 82. https://doi.org/10.1063/1.3632122

55. Gil-Santos E, Ramos D, Jana A et al (2009) Mass sensing based on deterministic and stochastic responses of elastically coupled nanocantilevers. Nano Lett 9:4122–4127. https://doi.org/10.1021/nl902350b

56. McFarland AW, Poggi M, a, Bottomley L a, Colton JS, (2005) Characterization of microcantilevers solely by frequency response acquisition. J Micromechanics Microengineering 15:785–791. https://doi.org/10.1088/0960-1317/15/4/016

57. Lee I, Lee J (2013) Measurement uncertainties in resonant characteristics of MEMS resonators. J Mech Sci Technol 27:491–500. https://doi.org/10.1007/s12206-012-1269-7

58. Kim S, Yi D, Passian A, Thundat T (2010) Observation of an anomalous mass effect in microcantilever-based biosensing caused by adsorbed DNA. Appl Phys Lett 96. https://doi.org/10.1063/1.3399234

59. Sader JE, Yousefi M, Friend JR (2014) Uncertainty in least-squares fits to the thermal noise spectra of nanomechanical resonators with applications to the atomic force microscope. Rev Sci Instrum 85. https://doi.org/10.1063/1.4864086

60. Naeli K, Brand O (2009) An iterative curve fitting method for accurate calculation of quality factors in resonators. Rev Sci Instrum 80. https://doi.org/10.1063/1.3115209

61. Van Eysden CA, Sader JE (2006) Resonant frequencies of a rectangular cantilever beam immersed in a fluid. J Appl Phys 100. https://doi.org/10.1063/1.2401053

62. Rabe U, Hirsekorn S, Reinstädtler M et al (2007) Influence of the cantilever holder on the vibrations of AFM cantilevers. Nanotechnology 18. https://doi.org/10.1088/0957-4484/18/4/044008

63. Tsuji T, Kobari K, Ide S, Yamanaka K (2007) Suppression of spurious vibration of cantilever in atomic force microscopy by enhancement of bending rigidity of cantilever chip substrate. Rev Sci Instrum 78. https://doi.org/10.1063/1.2793498

64. Malatkar P (2003) Nonlinear Vibrations of Cantilever Beams and Plates. Virginia Tech

65. Venstra WJ, Westra HJR, Van Der Zant HSJ (2011) Q-factor control of a microcantilever by mechanical sideband excitation. Appl Phys Lett 99. https://doi.org/10.1063/1.3650714

66. Marti O, Ruf A, Hipp M et al (1992) Mechanical and thermal effects of laser irradiation on force microscope cantilevers. Ultramicroscopy 42–44:345–350. https://doi.org/10.1016/0304-3991(92)90290-Z

67. Bircher BA, Duempelmann L, Lang HP et al (2013) Photothermal excitation of microcantilevers in liquid: Effect of the excitation laser position on temperature and vibrational amplitude. Micro Nano Lett 8:770–774. https://doi.org/10.1049/mnl.2013.0352

68. Evans DR, Tayati P, An H et al (2014) Laser actuation of cantilevers for picometre amplitude dynamic force microscopy. Sci Rep 4:1–7. https://doi.org/10.1038/srep05567

69. Sun J, Wu Y, Xi X et al (2017) Analysis of the damping characteristics of cylindrical resonators influenced by piezoelectric electrodes. Sensors (switzerland) 17. https://doi.org/10.3390/s17051017

70. Li W-C, Lin Y, Kim B, et al (2009) Quality factor enhancement in micromechanical resonators at cryogenic temperatures. In: TRANSDUCERS 2009 - 2009 International Solid-State Sensors, Actuators and Microsystems Conference. IEEE, pp 1445–1448

71. Kathel G (2016) Improving Mass Sensitivity and Reliability of Microcantilever for Bio-Sensing Application in Liquid Medium (MS Thesis). Indian Institute of Technology Madras, India

72. Mehdizadeh E, Chapin JC, Gonzales JM et al (2014) Microelectromechanical disk resonators for direct detection of liquid-phase analytes. Sensors Actuators A Phys 216:136–141. https://doi.org/10.1016/j.sna.2014.05.022

73. Mahajne S, Guetta D, Lulinsky S, et al (2014) Liquid Mass Sensing Using Resonating Microplates under Harsh Drop and Spray Conditions. 2014:

74. Arlett JL, Roukes ML (2010) Ultimate and practical limits of fluid-based mass detection with suspended microchannel resonators. J Appl Phys 108. https://doi.org/10.1063/1.3475151

75. Ekinci KL, Huang XMH, Roukes ML (2004) Ultrasensitive nanoelectromechanical mass detection. Appl Phys Lett 84:4469–4471. https://doi.org/10.1063/1.1755417
76. Ghatkesar MK, Barwich V, Braun T et al (2004) (2004) Real-time mass sensing by nanomechanical resonators in fluid. Proc IEEE Sensors 3:1060–1063. https://doi.org/10.1109/ICSENS.2004.1426357
77. Gil-Santos E, Baker C, Nguyen DT et al (2015) High-frequency nano-optomechanical disk resonators in liquids. Nat Nanotechnol 1–8. https://doi.org/10.1038/nnano.2015.160
78. Tappura K, Pekko P, Seppa H (2011) High-Q micromechanical resonators for mass sensing in dissipative media. J Micromechanics Microengineering 21:065002
79. Fukuyama M, Yamatogi S, Ding H et al (2010) Selective detection of antigen-antibody reaction using si ring optical resonators. Jpn J Appl Phys 49:1–2. https://doi.org/10.1143/JJAP.49.04DL09
80. Zhang W, Turner K (2004) A mass sensor based on parametric resonance. Proc Solid State Sensor, Actuator Microsyst Work, pp 49–52
81. Spletzer M, Raman A, Sumali H, Sullivan JP (2008) Highly sensitive mass detection and identification using vibration localization in coupled microcantilever arrays. Appl Phys Lett 92. https://doi.org/10.1063/1.2899634
82. Faegh S, Jalili N, Sridhar S (2013) A self-sensing piezoelectric microcantilever biosensor for detection of ultrasmall adsorbed masses: theory and experiments. Sensors (basel) 13:6089–6108. https://doi.org/10.3390/s130506089
83. Digilov RM, Abramovich H (2013) Flexural vibration test of a beam elastically restrained at one end: A new approach for Young's modulus determination. Adv Mater Sci Eng 2013:1–6. https://doi.org/10.1155/2013/329530
84. Surinėnaitė B, Bendikienė V, Juodka B (2009) The hydrolytic activity of Pseudomonas mendocina 3121-1 lipase. A Kinetic Study. Biologija 55:71–79. https://doi.org/10.2478/v10054-009-0012-5
85. David L (2008) CRC Handbook of Chemistry and Physics, 84th edn. CRC Press
86. Solanki PR, Dhand C, Kaushik A et al (2009) Nanostructured cerium oxide film for triglyceride sensor. Sensors Actuators, B Chem 141:551–556. https://doi.org/10.1016/j.snb.2009.05.034
87. Yücel A, Özcan HM, Sağıroğlu A (2016) A new multienzyme-type biosensor for triglyceride determination. Prep Biochem Biotechnol 46:78–84. https://doi.org/10.1080/10826068.2014.985833
88. Pundir CS, Aggarwal V (2017) Amperometric triglyceride bionanosensor based on nanoparticles of lipase, glycerol kinase, glycerol-3-phosphate oxidase. Anal Biochem 517:56–63. https://doi.org/10.1016/j.ab.2016.11.013
89. Brown LJ, Koza RA, Marshall L et al (2002) Lethal hypoglycemic ketosis and glyceroluria in mice lacking both the mitochondrial and the cytosolic glycerol phosphate dehydrogenases. J Biol Chem 277:32899–32904. https://doi.org/10.1074/jbc.M202409200
90. Valdés L, San José ME, Pose A et al (2010) Usefulness of triglyceride levels in pleural fluid. Lung 188:483–489. https://doi.org/10.1007/s00408-010-9261-4
91. Pramanik S, Harley K (2013) Evaluation of CSF-Cholestrol , Triglycerides and Electrolytes in Neurological Disorders. Ind Med Gaz 18–20
92. Thaler MA, Bietenbeck A, Schulz C, Luppa PB (2017) Establishment of triglyceride cut-off values to detect chylous ascites and pleural effusions. Clin Biochem 50:134–138. https://doi.org/10.1016/j.clinbiochem.2016.10.008
93. Fidal TV, Mottour Vinayagam C, Gayathri S, et al (2019) Detection of Urea and Ammonia with Aluminium Coated Polysilicon Nanoresonators. In: 2019 20th International Conference on Solid-State Sensors, Actuators and Microsystems and Eurosensors XXXIII, TRANSDUCERS 2019 and EUROSENSORS XXXIII. Institute of Electrical and Electronics Engineers Inc., pp 936–939
94. Lardi JM, George S, Township F (1994) Cleaning wafer substrates of metal contamination while maintaining wafer smoothness. US5498293A
95. Grubisic A, Li X, Gantefoer G et al (2009) Reactivity of aluminum cluster anions with ammonia: Selective etching of Al11- and Al12-. J Chem Phys 131:1–7. https://doi.org/10.1063/1.3256236

96. Hagendorfer H, Lienau K, Nishiwaki S et al (2014) Highly transparent and conductive ZnO: Al thin films from a low temperature aqueous solution approach. Adv Mater 26:632–636. https://doi.org/10.1002/adma.201303186

97. Gaddes DE, Demirel MC, Reeves WB, Tadigadapa S (2015) Remote calorimetric detection of urea via flow injection analysis. Analyst 140:8033–8040. https://doi.org/10.1039/C5AN01306B

98. Wei LF, Shih JS (2001) Fullerene-cryptand coated piezoelectric crystal urea sensor based on urease. Anal Chim Acta 437:77–85. https://doi.org/10.1016/S0003-2670(01)00941-2

99. Dutta D, Chandra S, Swain AK, Bahadur D (2014) SnO2 Quantum Dots-Reduced Graphene Oxide Composite for Enzyme-Free Ultrasensitive Electrochemical Detection of Urea. Anal Chem 86:5194–5921

100. Johansona CE, Stopa EG, Daiello L et al (2018) Disrupted Blood-CSF Barrier to Urea and Creatinine in Mild Cognitive Impairment and Alzheimer's Disease. J Alzheimer's Dis Park 08. https://doi.org/10.4172/2161-0460.1000435

101. Huang W, Bi Y, Hu Z (2014) Effects of fertilizer-urea on growth, photosynthetic activity and microcystins production of microcystis aeruginosa isolated from Dianchi Lake. Bull Environ Contam Toxicol 92:514–519. https://doi.org/10.1007/s00128-014-1217-6

102. Cao H, Li M, Hong Y, Gu JD (2011) Diversity and abundance of ammonia-oxidizing archaea and bacteria in polluted mangrove sediment. Syst Appl Microbiol 34:513–523. https://doi.org/10.1016/j.syapm.2010.11.023

Chapter 6
Conclusion

This chapter summarises the estimation of triglycerides and urea using an electro-chemical EISCAP and a mechanical resonant cantilever as sensors. Though both sensors are based on the enzymatic hydrolysis of the bioanalyte, the actual sensing mechanisms vary. These are compared, with an emphasis on the ease of measurement, sensitivity, reliability and the effects of miniaturization of the sensor structures.

We have presented an electrochemical sensor (EISCAP) and a mechanical sensor (microcantilever) used for the detection of two bioanalytes: triglyceride (TG) and urea. The basic sensing principle behind both the sensors is the enzymatic hydrolysis of the bioanalyte, which is a very specific reaction, though the detection mechanism is different—electrical in one case and mechanical in the other. The EISCAP detects the change in the pH due to the enzymatic hydrolysis of the bioanalyte, producing acid and bases, by measuring changes in the capacitance–voltage (CV) characteristics of the device. The microcantilever, on the other hand, measures changes in the resonance frequency as a consequence of the enzymatic reaction. The measurement mechanism directly affects the limit of detection, with the microcantilever achieving orders of magnitude lower limit of detection. Miniaturisation, using micromachining processes, improved the performance of the sensors.

The hydrolysis of TG in the presence of the enzyme lipase, produces fatty acid and glycerol as the reaction products. The EISCAP probes the change in the pH due to the formation of fatty acids while the microcantilever, immersed in the reaction chamber, senses the change in the density of its surrounding medium due to the formation of glycerol. Enzymes are like a prima-donna, requiring very precise ambience to work at the optimum activity, and these include the temperature and the pH. Optimisation necessitated the use of a buffer in the electrolyte, whose concentration should be low, since the buffer neutralises small changes in the pH at low concentrations of TG. The capacitance–voltage measurement for the EISCAP is found to be very sensitive to the buffer concentration, becoming unstable at low buffer strengths. Thus, for the measurements with the EISCAP, there is a tradeoff between the stability (and hence accuracy) and the lower limit of detection. The lowest value measured with the EISCAP was 0.56 mM of TG. which, luckily, is adequate for clinical investigations

© The Author(s), under exclusive license to Springer Nature Switzerland AG 2021
E. Bhattacharya, *Biosensing with Silicon*, SpringerBriefs in Materials,
https://doi.org/10.1007/978-3-030-92714-1_6

in the range of 50 to 150 mg/dL corresponding to 0.56–1.68 mM [1]. Since the microcantilever measurement is not dependent on the pH but on the density change due to the glycerol production, much lower limits of detection: 50 μM with the micro cantilever and 10 nM with the nano cantilever [2], could be achieved.

The readout circuit worked well with the EISCAP, simplifying the measurement protocol. A calibration step is essential for the EISCAP as there can be variations in the sensitivity with fabrication conditions [3, 4]. We have seen variation in pH sensitivity between 35 and 57 mV/pH among batch processed sensors. Immobilisation of the enzyme is convenient for the measurements, though it leads to a 20–25% drop in the activity as compared to the free enzyme. Enzyme can leach with time and an effective encapsulation protocol has to be developed to keep the enzyme active till it is time to use the sensor. The sensors can also work for other bio-analytes like glucose, with appropriate enzymes and change in the calibration and measurement algorithm to suit the specific application. An array of EISCAP sensors can simultaneously detect multiple bio-analytes at the same time. Automation of the delivery of the bioanalyte into a specific EISCAP can be achieved with the help of a delivery system like digital microfluidics [5].

The cantilever sensors give unprecedented sensitivity though there are issues with the ease of use and care has to be taken to ensure reliability of the measurement. Resonance frequency measurement in cantilevers usually require more sophisticated equipment like AFM or Doppler vibrometer. There are a few direct readout systems too that use either a feedback loop [6] or a PZT film [7]. But the highest sensitivity measurements reported usually involve a vibrometer. For ultrasensitive detection, the cantilever fabrication can also be complicated. But these sensors can play a very important role in specific applications, for eg, in detecting pathogens in early stages of a disease, when standard tests like the PCR take long enough for the disease to have already progressed rapidly.

In general, though there are many innovative biosensor devices and techniques for measurements developed in research laboratories, the number of these sensors making it into the market is still small. Perhaps a greater interaction with clinicians and involving industry in the early stages of development can mitigate this problem to some extent.

References

1. Veeramani MS, Shyam KP, Ratchagar NP et al (2014) Miniaturised silicon biosensors for the detection of triglyceride in blood serum. Anal Methods 6:1728–1735. https://doi.org/10.1039/c3ay42274g
2. Chinnamani MV, Bhadra P, Fidal VT et al (2020) Ultrasensitive detection of antigen–antibody interaction and triglycerides in liquid ambient using polysilicon cantilevers. J Micromechanics Microengineering 30. https://doi.org/10.1088/1361-6439/abb992
3. Garcia-Canton J, Merlos A, Baldi A (2006) A wireless potentiometric chemical sensor based on a low resistance enos capacitive structure. In: Proceedings of the IEEE International Conference on Micro Electro Mechanical Systems (MEMS), pp 462–465

4. Veeramani MS, Shyam P, Ratchagar NP et al (2013) A miniaturized pH sensor with an embedded counter electrode and a readout circuit. IEEE Sens J 13:1941–1948. https://doi.org/10.1109/JSEN.2013.2245032
5. Kumar P, Bhattacharya E (2013) Digital microfluidics and its Integration with a fluidic microreactor. J ISSS 2:10–19
6. Burg TP, Manalis SR (2003) Suspended microchannel resonators for biomolecular detection. Appl Phys Lett 83:2698. https://doi.org/10.1063/1.1611625
7. Lee Y, Lim G, Moon W (2006) A self-excited micro cantilever biosensor actuated by PZT using the mass micro balancing technique. Sens Actuat A Phys 130–131:105–110. https://doi.org/10.1016/J.SNA.2005.11.067

Printed in the United States
by Baker & Taylor Publisher Services